The Future and

Bitcoin

A Beginner Guide on Cryptocurrency

Mining and Blockchain Technologies,

Investing and Trading

By Elliott Dian, MASc

www.futurentech.com

Disclaimer

The material in this book or its affiliated website have no regard to the specific investment objectives, financial situation, or particular needs of any reader, user or visitor. This book is published solely for informational and entertainment purposes. No warranties of any kind are expressed or implied. Readers acknowledge that the author is not engaging in the rendering of legal, financial or professional advice. Please consult a licensed professional or licensed financial advisor before attempting any techniques outlined in this book.

By reading this book, the reader agrees that under no circumstances are is the author responsible for any losses, direct or indirect, which are incurred as a result of the use of information contained within this document, including, but not limited to, errors, omissions, or inaccuracies. The reader should use the information only as a starting point for doing additional independent research in order to allow you to form your own opinion regarding investment.

Table of Contents

Introduction

Bitcoin has taken the world by storm. It has captured the imagination of millions of people like a 21st century gold rush and it shows little signs of going away any time soon.

Despite government attempts to put blocks on it in countries like China and South Korea, and the inevitable scaremongering by mainstream media any time there is a significant drop in its value, Bitcoin has held on and continues to go from strength to strength.

You may wonder if you've missed the boat when it comes to investing, trading or mining for Bitcoin, or any other altcoin on the market, but that isn't the case. For shrewd investors there is still a lot of mileage in this and if you buy and sell at the right times you could still see huge profits going into your bank account.

This book was written to help new investors understand the intricacies, possibilities and risks associated with trading in any cryptocurrency. They are still in their infancy, decentralized, unregulated and pose significant risks for unwary or uneducated traders. But they are also exciting, thrilling and potentially limitless and you can easily get started with just a small amount of money.

Within these pages you can develop an understanding of how Bitcoin and other cryptocurrencies work, how to get started with your investment from just a few dollars and the best ways to trade, invest and store your cryptos.

For traders who have been in the crypto market and actively trading a lot, you may be starting to feel frustrated because the tide has changed a little in 2018. So this book is going to teach you how to find a quality setup. Once you learned how to find a quality setup, you can use the same strategy in the stock market as well, so you're not limited to cryptocurrency.

I will show you examples of some really good trades and will walk you through some exciting trading tools, including how to open an account and buy your first coins, how to link your account with trading exchange and how to take your first trades.

Let's get into it and start by talking about the $200 million pizzas. This is actually true; two pizzas sold for what would have been worth, in December 2017, $100 million each. That would be the story of Laszlo Hanyecz who back in 2010 purchased two pizzas for 10,000 bitcoins. Back then the value of those coins was practically $0. It was very speculative in the early days, but in just about less than eight years the value has skyrocketed to over 200 million at the peak of the bitcoin rise in 2017, when Bitcoin was at about $20,000. Those were the first transactions with Bitcoin and at the time there was no telling what it would someday become. At the time ten thousand Bitcoin were worth basically nothing, but it had this future potential.

Another thing that's worth noting is the fact that the Dow Jones industrial average took over a hundred years to go from around $30 to $20,000, and Bitcoin did it in less than a decade. It really goes to show the widespread consumer adoption of Bitcoin that we've seen in the last ten years.

This is a great place to get started on your investment journey and for a beginner to the world of Cryptocurrency it is essential reading, that will stand you in good stead and provide a foundation for all that you do for years to come.

If you have ever thought about investing some of your money in the cryptocurrency boom, then this is the book you've been waiting for. Read on and discover what is fast becoming the financial revolution of our times and be ready to claim your share of the cryptocurrency gold rush.

CHAPTER 1:

What is Bitcoin?

Bitcoin is a decentralized digital currency. In other words, it's similar to virtual cash or gold, which can be transferred person to person. This means there's no bank or government control or prerequisites to be able to own or transfer Bitcoins. It is a cryptocurrency, and it may surprise you to learn that cryptocurrencies have been around since the 1980s.

Initially, there was a major problem with the cryptocurrency concept, which was double spending. Double spending means the coins could be transferred to two or more people and it was very difficult to prove if they had already been spent. This was the biggest problem, but Bitcoin solved it and that is the reason it has gained so much popularity.

The first Bitcoin was issued in 2009 by someone named Satoshi Nakamoto. This is believed to be a fake name, so nobody really knows who started Bitcoin or if Satoshi is a real person. Some people believe that Bitcoin was created not just by one individual, but by a group of people.

Similar to gold or silver, there's a limited number of Bitcoins, capped at just 21 million. In 2009 we saw the first bitcoins released and there are a predetermined number of bitcoins that are released every 10 minutes, which I will talk about it later in this book. This will continue to happen until the year 2140. Bitcoin can be divisible down to the eighth decimal place (0.00000001). This is like the US dollar, which is divisible to two decimal places, or what we call a penny.

How does Bitcoin work?

Bitcoin is not designed on an account-based model. It is just a digital file that lists accounts and money like a public ledger, while also keeping track of transactions. The ledger is known as the blockchain. A copy of this file is maintained on every computer in the Bitcoin network. So, Bitcoin is a network of computers that maintain the blockchain. The accounts and values in the system don't represent anything in the physical world. They only have value because people believe they have value and are willing to trade with them.

To send money, account holders broadcast to the network how much they want to transfer from their own account to someone else's account. Computers in the Bitcoin network, called "nodes", apply that transaction to their copy of the ledger and then pass it on to other nodes. This is a system that lets a group of computers maintain a ledger. This is similar to the way banks maintain ledgers, but there are some differences as there's no central authority in Bitcoin to monitor transactions and accounts. In Bitcoin, everyone knows about everyone else's transactions. While you can trust your bank, or you can at least sue it if something goes wrong, with Bitcoin you are dealing with anonymous strangers.

Why should we trust this system then? The Bitcoin system is amazingly designed so that no trust is needed. Special mathematical functions protect every aspect of it.

Bitcoin is based on a pure peer-to-peer network. So, who runs the nodes in the network, and who provides computing power to keep the ledger or release new Bitcoins? Technically, anybody can run a bitcoin node to verify transactions. The people or computers that are providing computing power are called miners. Anyone with access to the internet and suitable hardware can participate in mining.

One of the big advantages of Bitcoin is that nodes don't have identities. In a decentralized peer-to-peer system, there's no centralized authority to verify identities and ensure each identity is not creating another node. Just keep in mind, Bitcoin doesn't give you any guarantee that your identity can always be hidden, because your identity can be found out based on the series of your transactions linked together. But at the same time, the Bitcoin system doesn't require your identity such as your name or IP address to be a part of the network. This is an important property.

Bitcoin is not backed by gold. It's not stock either. It can be traded like stocks, but it has its own exchanges. There are no brokers for exchanging Bitcoin, which can be a big advantage if you're a trader.

What is a Bitcoin account?

There's actually two parts to your Bitcoin account. Number one is your public key which is similar to your any account number or your email address. A Bitcoin key is made up of 27 to 34 alphanumeric characters. This is an account number that you can give out to anybody who wants to send you bitcoins. The other part of Bitcoin is your private key. This is similar to your password for any online accounts. With your private key, you can access your Bitcoins in the account.

There are a few ways to acquire Bitcoins. One is buying directly from a person, but you can also go to an exchange, like a stock market, and trade in it. The other way is mining for Bitcoins. I'll explain these methods in more details in the coming chapters.

How can you profit with Bitcoin?

One simple way to start is mining. If you have a computer system at home, you can mine and earn Bitcoins, then sell them for whatever the market value is.

The second way is to trade or invest in Bitcoins. You have probably heard a lot of people say that Bitcoin is a bubble. There's a lot of speculation. Prices have increased and also dropped dramatically over the past couple of years. But keep in mind that fluctuation in the price is actually a positive thing in trading. You can buy and hold Bitcoins, but it could be risky. Some people have made vast amounts of money out of it, but some others have lost a lot too. So, it could be very risky. A better approach in a fluctuating market is short-term trading. This method has less risk and there's opportunity to make profitable trades in this space.

A lot of people and companies are accepting Bitcoins instead of money for their services too. So, this could be another way of earning Bitcoins. Recently, a lot of big online merchants have started accepting Bitcoin.

CHAPTER 2:

Benefits of cryptocurrency

Why is Bitcoin important?

Earlier in the previous chapter, we talked about how Bitcoin works without a centralized authority and without any controls from the government or financial institutions. This major property of all cryptocurrencies can introduce some risks, but at the same time provides a lot of big advantages. There are some aspects of cryptocurrencies that show huge prospects for dramatically changing the world we live in.

Out of the seven plus billion people that live on the planet, about two billion people do not have a bank account or access to basic financial and banking services. Bitcoin makes it very easy for anybody without government permission to have access to virtual banking. This is really going to change the financial prospects for these people.

We are seeing big inflations in Fiat currencies. And just to remind you, they are backed by nothing but the trust that people have in their government. Back in the 70s we were on the gold standard, but now fiat currencies like the US dollar are not really tied to anything. We can also look at the Zimbabwean dollar as kind of a reminder of what can happen when governments get out of control by printing too much money. In the span of almost two years, the Zimbabwean government essentially devalued their currency to nothing. Bitcoin can be used as a global currency when we can no longer trust our fiat currencies.

The real value of Bitcoin is determined by what people are willing to pay for it. The increase in people's trust in the system has led to increase in demand. This is the reason we've seen the price of Bitcoin go up so much over the past couple of years.

Cross-currency purchases and transfers across borders are expensive and have seen a lot of friction and government red tape. If you want to move millions of dollars, it would cost a lot of money and would take a long time to happen. Where with Bitcoin, it's inexpensive and fast. We can instantly transfer unlimited amounts of money with very low fees. Also, when you transfer money between countries, you won't lose out because of currency exchanges.

It's very easy to buy and sell with extremely low transaction fees. Sending and receiving Bitcoin is as easy as sending an email.

It is also transparent; it means the code in the structure and how the Bitcoin cryptocurrency network and protocol is built is there for all to see. Anybody can look at the code, see how this is built and evaluate it. Also, its public ledger cannot be removed or changed.

Cryptocurrency eliminates the middle man. It's not under the control of a central authority. People place value in it because there's a lack of government control and manipulation. So, it's protected from inflation.

Cryptocurrencies can be used in any transactions; from big ones like real estate, to small purchases like food. Companies from different industries are beginning to accept Bitcoin as a legitimate source of payment. Some of these are well-known companies like Microsoft, Subway, Square, Shopify, Newegg, Expedia and Virgin Galactic. And the list is getting longer.

The way Bitcoin protocol and network is designed makes it very difficult and expensive to hack. It's controlled by the consensus of market participants. That makes the network trustful.

CHAPTER 3:

The future of Bitcoin and cryptocurrencies

Is Bitcoin a bubble?

Bitcoin has increased in value since 2009, from a few cents to about $19,000. Through the rise in price, we've also seen many crashes. But prices have mainly recovered after each crash. What will happen to Bitcoin and all other cryptocurrencies in future? Nobody can really answer this question.

In 2017, the price of Bitcoin went up about 2000% and then right after, it had a pretty big crash. This raised a question for a lot of people about what is going to happen next. Over the past few years, Bitcoin has gone through several booms and bust price cycles. But generally over time, the price of Bitcoin has just kept going up. If you understand the way these cycles work, your investments can be profitable compared to the majority of people that let their emotions dictate when to buy and sell. Fear of missing out causes people to follow big booms and parabolic trends, and then they panic and sell during crashes. The media and news make this problem even worse. Because when prices in a market are going up, the media loves to put out all these positive articles about it, and when the prices are going down, they do the exact opposite. Just keep in mind that media just want to release articles that are sensational.

As a trader, I'm a huge proponent of not chasing what the herd is doing. We saw this in the housing bubble of the early 2000s when people got excited about investing in real estate. That drove prices up, and then it crashed. The idea is to understand that you're investing in something that is going to have long-term value.

To better understand the Bitcoin market, let's look at some previous bubbles in the price of Bitcoin. The first bubble we saw was in June 2011, where Bitcoin price went from about 50 cents to $32 in about 60 days. It crashed back to 2 dollars over the course of a few months. This is when the first enthusiasm for Bitcoin for the majority of people came in. In April 2013 we saw another big bubble where we went from about $35 up to $266. It again crashed back to $50 in the span of a few days. That then recovered and saw the next bubble of November 2013, where price went from a $100 up to $1200 and then crashed below $400.

It's important to keep in mind that bubbles are relative, meaning if we look at these past bubbles in 2011 and 2013, you will notice that, in hindsight, after the big spike in 2017 these little bubbles just seem like unimportant blips on the chart, as shown in the picture below.

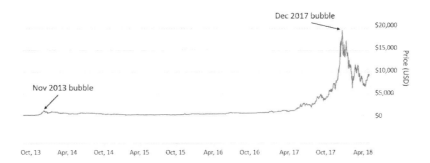

Bitcoin bubbles

When there's unsustainable and dramatic price increases in almost any market, such as gold, fiat currencies or stocks, we see big crashes after them. When the prices are going straight up, everybody can follow the rule of buying low and selling high and make money. But when the market crashes, or it's volatile, that's what separates the real traders from emotional investors. When we see fluctuations in a market like Bitcoin, we have good opportunities for good short trades, some good long swing trades or even day trades. I'll explain more about how trading Bitcoins and other altcoins in the following chapters.

CHAPTER 4:

Alternative to Bitcoin (Altcoins) and ICOs

Altcoins

In this section I am going to review some of the different cryptocurrencies that are available and explain about ICOs. Altcoins are also known as Alts and are cryptocurrencies that are not Bitcoin. The word Altcoin is an abbreviation of alternative coins. Some well-known altcoins are Litecoin, Ripple, Dash and Ethereum. All other non-Bitcoin cryptocurrencies also fall under the category of being an altcoin.

There are over a 1000 cryptocurrencies in existence. This picture here shows a number of different cryptocurrencies. One of the things that you'll notice is that they're all very cool little icon type of shapes. There are more cryptocurrencies than you could remember.

Cryptocurrencies icon shapes

Why do we even need Altcoins in the first place? The answer is simple; Bitcoin is not perfect. Usually Altcoins try to create a different or better version of Bitcoin. For example, Litecoin is a type of altcoin that confirms transactions faster than Bitcoin. Dash focuses on the anonymity aspect, making transactions virtually impossible to trace. Each Altcoin has its own unique feature. Altcoins can vary from Bitcoin in the way they are mined. For example, the Bitcoin mining algorithm is called SHA-256 while the Litecoin algorithm is called Scrypt. Different mining algorithms require different hardware to mine. Another thing to keep in mind is that if an altcoin is relatively new, or not that well known, it'll probably be harder to buy and will have fewer wallets that's supported. At the time of writing, thousands of altcoins have been introduced to market. But only a few of them have managed to get a significant following like Bitcoin has.

How to choose altcoins to invest in?

One way to figure out which altcoins are gaining popularity is by measuring their market cap. Market cap (short for capitalization) means how much money or capital is invested in an asset when measured in dollars. It's calculated by taking the number of coins in circulation and multiplying them by the dollar exchange rate. If, for example, there are 10 million Bitcoins in circulation, and the price of one Bitcoin is 2500 dollars, then the market cap for Bitcoin will be about 25 billion dollars. For a long time, Bitcoin's market cap accounted for 90% of the total cryptocurrency market cap. Today though, as more altcoins are gaining attention and appreciating in price, Bitcoin's share of market cap is going down in the total cryptocurrency market cap, which managed to surpass a hundred billion dollars in 2017.

How do you decide in which altcoins to invest in? First you need to read about the altcoin you're interested in and make sure you understand what makes it unique. Most importantly, don't invest in a coin just because of the hype. Many coins out there are known as a pump and dump scheme. This means that the coins creators generate a lot of hype about a specific coin in order to get people to invest in it and then inflate the coin's price. Once the coins appreciate in value, the creators sell all of their coins and profit, while crashing the price due to the massive sell off. This leaves the majority of investors with a bunch of useless coins and no one to sell them to.

I would also suggest getting involved in the coin's community. Usually most major coins have an official forum, where you can ask specific questions from the developers of the coin. A strong community is an important predictor for a coin's success.

ICOs

ICO stands for Initial Coin Offering. The term comes from the traditional finance term, IPO or Initial Public Offering. An IPO is used to describe the launch of a new company on a stock exchange. Also known as going public. The purpose of an IPO is to sell stock in a company in order to raise capital from the public. ICOs, on the other hand, sell coins, also known as tokens, as a way to fund a specific project. The general idea is that if you believe a project will succeed, you buy the tokens that power the projects beforehand at a discount. And then you'll be able to sell them later at a profit, if and when the project succeeds.

When a cryptocurrency company wants to launch a new project through an ICO, it creates a whitepaper. The whitepaper is a document that states what the project is about, what needs the project will fulfill, how much money is needed for the project and how long the ICO for this project will run for. After the ICO is set up, the public starts investing in the ICO by sending money to the projects and receiving tokens or coins in return. If the money raised by ICO does not meet the minimum funds required by the whitepaper, the money is returned to the bankers and the ICO is deemed to be unsuccessful. If the fund requirements are met within the specified timeframe, the founding team will get to work and bring the project to life through the use of the funds raised.

ICOs are like Kickstarter for crypto projects. The best example of a highly lucrative ICO was the presale of Ethereum tokens. In mid-2014, one Ethereum token, also known as ETH, sold for around 40 US cents. If you bought a hundred dollars of ETH, you would have the equivalent of 75,000 dollars in 2017. Today, more and more projects are trying to mimic the success of the Ethereum ICO.

ICOs are conducted over the Ethereum platform. That's why you need to buy the Ethereum tokens to be able to participate in them. The recent high volume of ICOs caused Ethereum's price to spike and overwhelmed the Ethereum network. This caused delayed or failed transactions, leading to the suspensions of

Ethereum trading on several exchanges, and problems with ICO funding.

ICOs can be very profitable, but at the same time contain a huge amount of risk. The worst example of a disastrous ICO is the DAO. The DAO, or Decentralized Autonomous Organization project, managed to raise a 150 million dollars' worth of Ethereum. However shortly after the ICO ended, a hacker managed to drain a third of the amount raised, due to a bug in the Ethereum code. This crisis, and different opinions on how to handle it, led to a split in the Ethereum network, and the creation of both the Ethereum and Ethereum classic altcoins.

ICOs can be considered as high-risk gambles on cryptocurrency startup companies. Many people today invest in ICOs, not because they believe in the project, but because they just want to make a quick profit. This creates a general hype before the project launches. But as the buzz fades away, project creators and early investors want to take the money off the table. So, they start selling massive amounts of tokens and this can cause the price to drop sharply.

Another thing to consider is that the bar for creating an ICO today is pretty low. While conducting an IPO requires a lot of regulation, ICO skips this entire burdensome procedure by raising money exclusively in cryptocurrency which has yet to be regulated. Want to create an ICO? Just create a shiny new website outlining your concept, create a digital asset, get some crypto celebs to say nice things about you and sell your project assets directly to people around the web. You don't even have to have a working product.

These low barriers to entry bring about a mass of low quality projects that will never see the light of day. This is the scammers dream come true. With minimum investment in time and money, they can get thousands of people to send them money without any legal exposure or liability. Sometimes you can lose money in ICO, not due to an intended scam, but due to a hacker manipulation. For example, not long ago a hacker managed to get into an ICO website and change the Ethereum deposit address to his own address. Over 7 million dollars was sent to him. That money is now lost forever.

How to choose ICOs to invest in?

Due to the irreversible and unregulated nature of cryptocurrency, you have virtually no recourse if an ICO turns out to be a complete scam. How do you know if you should invest in an ICO? You need to read the ICO's whitepaper, research the project and founders and get involved with the community around it.

Make sure you understand how the tokens for the project will be distributed. ICOs which hold a big share of tokens for the founders may end up selling these tokens in order to make a quick profit after the ICO ends. Another important question to ask is how much money is being raised and for what purpose. If there's no cap on the amount being raised, the project may get overfunded. Most importantly, never invest in something you don't understand. Do lots of research before committing your money to any project and understand that ICO is a new form of crowdfunding, which only a few people really understand.

CHAPTER 5:

Trading and investing in Bitcoin

Trading between coins is similar in many ways to trading between fiat currencies like dollars or euros. The coin prices fluctuate back and forth, depending on the demand or how interested people are in them.

To explain trading in a simple language, there are two common ways to make money through trading. One is going long, which is buying low and selling high. You can use that principle with stocks and real estate, and now you can do it with Bitcoin and other Cryptocurrencies. You can buy something for $1 and sell it for $5 and keep the difference. What a lot of people don't know is that you can actually make money as the price of things goes down as well. When you go short, you actually sell first. You sell high, and then buy back later at a lower price and keep the difference. This way you can make profit as the price of something goes up and as the price of something comes down. You can't short everything; as an example, you can't short physical real estate. But you can short stocks, ETFs and you can most certainly short Bitcoin.

Another way is Arbitrage, where you buy something, let's say a Bitcoin, at exchange number one and you immediately sell that same Bitcoin at Exchange number two and keep the difference. Arbitrage is very difficult to do in the Bitcoin market, as the fees are quite high. So, in a lot of cases it doesn't really make sense to do it. What I recommend is directional trading where you go long or short. Here in this chapter, I'm going to teach you a simple strategy to trade.

Before we start, I want you guys to understand the risk involved; obviously trading stocks is risky, and trading in cryptocurrencies is very risky too. I'm going to break down some of the levels of risk that you really need to be aware of before you get into this. I will also talk about the critical concepts that you need to know. These will help you become more profitable sooner and help you avoid the common mistakes that many new traders make.

Why trade Bitcoin?

I would say the big opportunity here is that when we look at cryptocurrencies, we see the patterns that are used on stocks work out in the same way and resolve in the same way. What really makes crypto different is that while in the stock market we see some big moves in some stocks, the crypto parabolic moves are much bigger. In crypto space, Bitcoin went from 1,000 to 20,000. We will look at charts for Ethereum and Litecoin, and you will see these have equally large parabolic moves.

The prices of Bitcoin or other alternative coins are extremely volatile, meaning there are big swings in price. It may look scary, but for traders, whether day traders or swing traders, that's where the money is made. When things are kind of flat and steady, it's more difficult to pull money out of a market. What creates this volatility right now is that there's a lot of emotion in Bitcoin and other cryptocurrencies. There's a lot of speculation and a lot of people that are just finding out about them. If you're a professional trader and you understand human psychology, you can make a big profit in a relatively short period of time.

When you are trading stocks and you buy $500 of them, you maybe take that $500 of risk and you make $500 or maybe $1000. If you take that same level of risk in cryptocurrencies, you could make four times, five times or six times in return. So you'd risk $500 and you could make $5,000 or $4,000. It's just because of this incredible excitement.

Another benefit, and definitely an appeal of cryptocurrencies, is that you can trade 24 hours, including weekends, and there's no minimum trading balance required. The risk you need to consider in trading or buying crypto is that right now there's not a lot of regulation. In the stock market there's a lot of regulation about when companies can do IPOs; what you can buy; what you can sell; what you can short. With the cryptocurrency, it's a bit like the Wild West right now, which has caused some of the big drops. Because when investors read a headline of possible regulation by a country, for example a country says we're going to regulate or no one is allowed to mine cryptocurrencies anymore, all of a sudden that can have a really drastic impact on the value. That's a risk component that you have to be aware of.

But right now, no minimum trading balance with no regulation is a plus if you only have a couple of hundred dollars that you want to throw into it. It's a great place to be involved as a trader. If you're still working a nine-to-five or for any other reasons you can't trade the stock market, you could be applying same trading strategies to cryptos and honing your skill, building up a little bit of confidence and then transition to trading stocks once you retire from the nine-to-five.

One of the things that is also interesting is that in the last few months we've seen a number of stocks informing us that they were developing blockchain technology and suddenly we saw these stocks just take off. We saw one stock go from $6 to $155 a share in four days, which is absolutely crazy. It was one of the biggest moves I've ever seen, and it was definitely because of what was happening in the crypto space. That headline got traders excited and I guess for traders like me that don't really trade or haven't traded at that time, cryptocurrencies was an opportunity to be part of the excitement. That's definitely something that is worth noting.

If you're trading an altcoin, remember Bitcoin can go down to the one 100 millionth of a Bitcoin, so if you own it at basically zero and it does go up to 0.0000001 you're still at a very good profit. So you'll see a lot of these trades at far less than a penny, and these slight little moves for the people that are in it really early could mean a big profit. You just have to be mindful that the pump-and-dump stuff that we've seen with penny stocks has happened to a certain extent in the crypto space as well.

Institutional adoption

Institutional adoption is an interesting topic, because initially a lot of Wall Street investors were super critical of cryptocurrencies. At the beginning, they all believed this was a bubble and then they were forced to start respecting it as the price got higher and higher and more people were adopting it. We now are seeing Bitcoin futures. It's an unusual situation because usually the institutions are the first to the party and the retail investors are last, but this has totally flipped in this situation where the retail investors got in first on the cryptocurrencies and it's the institutions that are the late comers.

We saw in December of 2017 that the CME Group and the CBOE group both rolled out Bitcoin futures. We also heard that NASDAQ plans to roll out Bitcoin futures in 2018. These are some of the most reputable financial exchanges in the world and they're now getting involved with Bitcoin. Financial institutions were really unable to be involved with Bitcoin prior to Bitcoin futures because of the unregulated nature of the Bitcoin market. But now with the futures markets, those regulations are in place; that opens the door for institutional inflows. It also allows for the shorting of Bitcoin which gives traders the ability to profit to the downside if they think Bitcoin is going to decline in the future. This was a challenge for retail traders, because they didn't have the ability to short Bitcoin. So you could only really bet on it going one direction and obviously that has probably propelled the move up because you didn't have any choices, even if you wanted them, to take the opposite side. It just wasn't an option. Financial institutions have to trade in regulated financial instruments when they're trading the money of companies like 401k or pension fund. They can't just be dabbling in an unregulated market like cryptocurrency. But now with the futures, they're able to do this.

Making profit by trading Bitcoin

To get into this, let's start looking at the charts of major cryptos here.

Bitcoin:

The parabolic Bitcoin move in 2017, we see in the stock market all the time. Traders get so excited, they see opportunities and stocks go kind of parabolic, they go almost straight up. But if the fundamentals of the company in the stock market can't sustain those values or justify them, they do come down. In the crypto world, this value is based on what people perceive it to be. So there was a bit of a profit taking opportunity in 2017 and then consolidation in the year after. It will be very interesting to see if this just fades back down to the 2,000 or 3,000 dollars level, or if it curls back up. At this point, it's anyone's guess and we'll find out in due course.

Ethereum:

Ethereum is one of the major cryptocurrencies that practically exploded in value in 2017. Ethereum has a supply of 96 million coins. With the price of $800 per coin, that gives it a market value of 78 billion dollars.

Ethereum went up nearly 10,000% since its low in the beginning of 2017, which was around $8. To put that in perspective for you; every $100 invested at the beginning of 2017 would now be worth $10,000.

One of the reasons why Ethereum has been one of the leading cryptocurrencies is because it provides people with the ability to make their own. The reason why we've seen so many more come on to the market is because platforms like Ethereum have made it so easy for anybody to come up with their own cryptocurrency.

To get more into the fundamentals of Ethereum, it is actually a decentralized blockchain based platform that allows developers to build their own decentralized applications. So users can build their own cryptocurrencies using the platform and that has paved a way for the explosion in the number of total cryptocurrencies out there. They use a technology called a smart contract which helps to verify and enforce contracts without the need for a centralized third party. Right now the contracts are still in their early stages of development, but there are numerous applications for them.

You can think Ethereum to be the racetrack while all the cryptocurrencies that are built on the platform would be the race cars. One of the reasons why Ethereum has gained so much in value is because the market realizes that it's better to bet on the platform that deploys all of the cryptocurrencies, rather than betting on the individual cryptocurrencies themselves.

Litecoin:

Total supply of Litecoin is 84 million. Litecoin can be thought of as a competitor to Bitcoin because it is attempting to solve all of Bitcoin's major flaws. For example, one of the major upgrades of Litecoin is that it can transact much faster than Bitcoin. While it takes Bitcoin about 10 minutes to generate a block, it takes only 2.5 minutes on the Litecoin network. It can also handle a much higher volume of transactions than Bitcoin, which is the result of the faster transactions.

If we look at the Litecoin chart here you can see this crazy move from down under $60 around October 2017, all the way up to

$380 in just two months. But again, like Bitcoin and Ethereum, it had a pretty steady pullback.

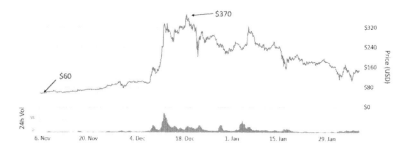

Litecoin parabolic move

What you can see here, and I'll talk about this a little more when we're breaking down the patterns, is that this right here is one of the patterns that we buy. When we see a stock start to breakout, we wait for a couple red candles and then the first candle that breaks the high of the previous candle is our entry. You can see here your entry would have been right under a hundred bucks and this just went basically straight to $380. This is a crazy return and our max loss is to the low of the pullback. Later in this chapter I'll explain more about this trading strategy.

These kinds of trades are $500 of risk for $10,000 profit with only $500. You can imagine traders who took $5,000 of risk or even $10,000, they were just destroying the market.

Litecoin is pretty unique because it has had a crazy breakout. Most of the other cryptocurrencies took some time to travel but this thing literally, in just three or four days, went absolutely parabolic.

Ripple:

Ripple is a payment network and a cryptocurrency that was created in 2012 to connect large financial institutions such as banks. It allows for the transfer of money through what's called the ripple net which is the payment network. XRP is the cryptocurrency that is used within the payment network. Transactions settle in four seconds, whereas with Bitcoin it can take over an hour. Ripple can also handle up to fifteen hundred transactions per second, whereas Bitcoin only does about seven. This is a major upgrade in infrastructure over Bitcoin.

Ripple (XRP) is another crypto that had the most incredible moves out of the other cryptocurrencies in December of 2017 when it was as low as 30 cents and then all the way to $330 in just under a month. It may seem like it's a cheap cryptocurrency, but it has a really large supply and it was valued at about a 127 billion dollars at the highest of $330. It again retraced much of the value and it is back down to about 75 cents. But that's not alarming considering how steep and quick that rally was.

We have seen this type of price action in stocks too; stocks that go from 30 cents or 40 cents to $3. They come back down and a lot of times they never make the way back up. It will be really interesting to see if cryptocurrencies are able to hold their value and if consumers are willing to step in and buy them at some point.

Four trading concepts

Let's talk about the trading concepts that you should master before you start trading. The good thing about cryptos is that you can start with a very small amount of money, and there are no monthly fees on the exchanges. This is very important because you'll find that as a new trader you're going to make beginner's mistakes and that's okay. You want to make those mistakes with small amounts of money. In order to be successful, you need to master these four concepts:

1) Risk management
2) Fundamental analysis
3) Technical analysis
4) Trade execution

In mid-2017, a lot of traders got into the market because they got on the crypto bandwagon, now cryptocurrencies are slowing down and they're wondering what to do next. It was the same thing a couple years ago when a lot of traders got into the market buying the marijuana stocks because they were going crazy with potential legalization, but then that market slowed down and those traders got into trading other stocks.

What is interesting about these concepts is that they are universal market concepts for profitability; they can be applied to practically any markets such as stocks, cryptocurrencies, Forex and commodities. Mastering those four concepts are critical to ensuring your success as a trader.

1) Risk management

The single most important concept for achieving consistent profitability would be risk management. You can't get involved in any trade without first understanding what your risk is, or what you stand to lose if you're wrong.

You have to realize with cryptocurrencies, probably more than other financial instruments, there's a risk of losing not just a lot of your money but potentially all of it. Right now cryptocurrencies are unregulated that there is a risk that something could potentially happen. So, you must be mindful of that.

Also recognize that cryptocurrencies are volatile, and they can have large spreads. What this means is that as soon as you buy the crypto, you're buying from someone who's willing to sell it to you. The distance between the price of a buyer and the price of a seller is the spread. If you buy Ethereum at $700, you're buying it from someone selling at $700. But let's say the nearest person willing to buy it is only at $699 that's a $1 spread. Now suddenly if you turn around and want to sell it a minute later you just lost $1 per share which could be significant depending on your position size. You have to be a little careful there before you take the trade. The cryptos on Coinbase exchange have tighter spreads, so the risk will be less. I'll explain how to buy and sell on Coinbase later in this chapter.

As a trader your number one priority is to keep your losses as small as possible. I recommend you don't risk more than one percent of your account in any single trade. For example, if you have $10,000 in your account, you shouldn't be willing to lose more than $100 on any single trade idea.

On the other hand, you should also be aiming for a minimum risk reward ratio of 1 to 2 on every trade. This means that you want to make at least 2 times what you've risked; if you're willing to risk $100 on a trade, you should have the ability to make at least $200. If that potential doesn't exist, then it's a trade that's not even worth taking.

This is the table on the profit to loss ratio. When you risk a hundred to make a hundred, you only need to be right 50% of the time to break-even, which is great because you've set the bar low. But what would be fantastic is that if you risk $100 to make two hundred, which is a two to one profit loss ratio, the bar is even lower. You only need to be right 33% of the time to break even. Then if you're right half the time you're actually really profitable. If you're right 60% of the time, you're doing extremely well.

REWARD	RISK	BREAKEVEN WIN RATE %
1	50	98%
1	10	91%
1	5	83%
1	3	75%
1	2	67%
1	1	50%
2	1	33%
3	1	25%
5	1	17%
10	1	9%
50	1	2%

Risk to reward ratio

The thing is that a lot of new traders don't even know about profit loss ratios and risk management. So they end up taking trades where they risk three, four or five percent of their account, or they risk $500 to make only a hundred. At the beginning, do everything you can to keep your losses as small as possible. This way, you can survive the learning curve to get through to the other side where you can start taking more risk, gaining more confidence and building more profits.

2) Fundamental analysis

This is one of the other core concepts that traders really need to understand. It mainly helps on long term investments. Fundamental analysis is different with cryptocurrencies than it is for trading stocks. When we think about fundamentals in stocks, we think about the company's earnings reports, their quarterly earnings, their annual earnings, press releases, the industry they're in and so on. These are fundamentals that we look at, more as an investor than as a trader. As a trader we focus on trading just the price. So we don't have to put a lot of emphasis into doing fundamental analysis. When you're doing a trade that's as short as 10 minutes, 15-20 minutes or even a couple hours, long-term fundamentals are not going to impact your trade. What impacts your trade is just that short term volatility. When it comes to fundamentals on cryptos, you do have to be mindful of the fact that we have a bit of a risk with lack of regulation. There's certainly headlines that come out about blockchain or ICOs that you need to know about and analyze to make some decisions.

3) Technical analysis

The next component is technical analysis which is the essential part for short term trading. We analyze the price to make predictions about what the price is going to do a few minutes from now, an hour from now, two hours from now or a week from now. That's a short-term type of technical analysis.

What we use are Candlestick charts and we use a small handful of technical indicators including moving averages and volume bars. The way to use these is to look for indicators of potential breakouts. When we see an indicator of a potential breakout we often buy either just before the breakout begins or right after. Basically we look for the cryptos that are starting to move. When we have an imbalance between supply and demand, that's when you have home run potential and you see big moves in cryptos or stocks.

One important question here is how to find right entry points to trade? Where to get in and where to get out? Here I'll teach you one powerful strategy. As soon as you learn this one strategy, you can become a successful trader.

If you're a beginner in trading, what I recommend you is keeping it simple. Get rid of multiple indicators, and just start with basics. Here's my simple strategy; when we're analyzing candlestick charts, what we look for are periods where a chart is showing a moment of rest. The chart below shows a moment of rest; we hit a new high and we've pulled back and consolidated.

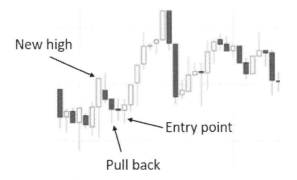

New high

Entry point

Pull back

Trading entry point (empty candlesticks represent green, filled candlesticks represent red)

Whenever we have a few minutes of consolidation, this is an opportunity for an entry point on a pullback. So, I like to see candles that are squeezing up (big green candles) and then I like to buy in the dip. Sometimes the dip (red candles) is two candles, sometimes it's four, sometimes it's only one and that's okay. In the picture below, we see a big green candle on the left side and then a few red candles after, this is the dip I'm talking about. As soon as a green candle shows up (the fifth candle from the left) and passes the high of the previous red candle, we start buying.

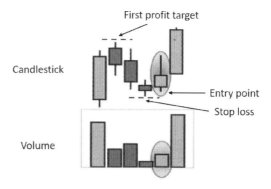

Breakout pattern (first candlestick from left and 2 first candlesticks from right represent green, the rest are red)

Here the main question is where to put our max loss and profit target to minimize the risk? In trading stocks or cryptos, loss is part of the game, and we can't avoid it. All we need to do is minimize those losses.

For the max loss, always put it at the low of the pullback. When we buy the green candle after pullback and the price goes down instead of going up, the low of the third candle stick is where we should stop. On the other hand if the price goes up, first profit target is that we hit the high of the day. The second profit target is that we squeeze up over the high of the day, and then from there you never know. They can go straight up, or they can go up and then pull back. So my max loss is the low of the pullback, profit target is the high.

You can use this strategy on a five-minute chart, one-minute chart or you can even trade it on a daily chart as a swing trader.

This is a universal pattern; you can trade this pattern on cryptos, stocks, futures, Forex or on any exchange in the world. This is a respected language; it's like sign languages that works all around the world.

Here's another point that's really important. You may see these types of patterns quite often. But they work in only certain situations; you can only trade this pattern on the right cryptos or stocks that have high volume and are moving quickly. You need to be trading those for one reason; those are the cryptos that all the other traders in the world are also watching. The fact that they're watching it is what creates the volume, and that's what creates the breakout. So that's what we look for which is so important.

What really allows you to predict future direction of prices is simply by studying what has happened in the past. You can find so many of these patterns every day in crypto space or stocks. Here I'll provide one example.

The point I indicated in the picture below is the beginning of a breakout.

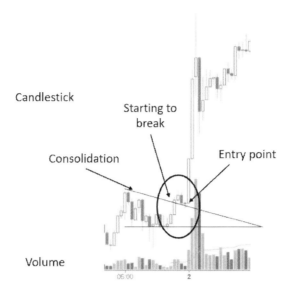

Breakout pattern in Ethereum (empty candlesticks represent green, filled candlesticks represent red)

These four candles are indicating the crypto (in this case Ethereum) that is starting to break. So our entry point is the first little pullback. We call it a micro pullback. We buy the pullback and then ride it for the next leg up.

Let's look at the 1-hour chart of Ethereum. This is the chart that we were just looking at before in previous picture. Section A in the picture is the period of consolidation and we didn't know when it was going to breakout. But boom, it started to break out at point B, it pulled back, and we bought at micro pullback. That was an entry right around $750, which never looked back.

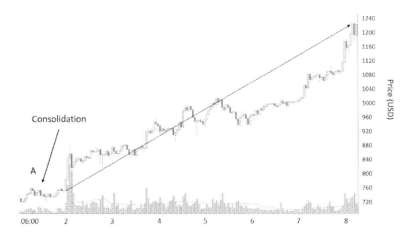

ETH 1 hour chart (empty candlesticks represent green, filled candlesticks represent red)

One of the strategies that we use for position management is called sell half and hold the rest with a stop at breakeven. Breakeven is the price that you got in at. When the price is going up, we would often sell half into a sharp move to make some profit and then hold the rest as long as possible. As the price is going up, you could continue to sell the other half in small portions and slowly sell it off.

This could have been a huge home run trade for a beginner trader. You don't see moves like this in the overall market very often, but in the crypto space you're seeing them happening frequently. Even if it turns out that the crypto space and its whole upward move was totally a bubble, it doesn't mean that you couldn't have been very profitable in it.

The lessons that you learn here, with what happened during this crypto bubble, you can apply when we see another one. We saw many in the stock market a couple of years ago; when we had the Ebola crisis, any company that made anything even close to an Ebola vaccine was going up like three or four hundred percent. That was because there was the possibility of it becoming widespread. So those companies became valuable. Another example is the move in cannabis companies during legalizations. We see these types of bubbles and whether they're in the stock market, entire market or just a sub sector of the market, they are opportunities that technical traders can capitalize on.

4) Trade execution

The next core concept is trade execution. For executing trades, you have to first perform your risk analysis and at least your due diligence on fundamental and technical analysis. Once you have broken down the pattern and you're confident that this candle is about to make a new high, now it's time to be ready to press the buy button. The important thing here is to remember to be focused on discipline. It's so easy as a trader, if you get frustrated, just to keep pressing the buy button to buy more and more. So, there's a process of emotional conditioning. To avoid that, risk-taking should be done in a calculated way; by taking time and measuring our risk we reduce how much we're willing to lose, which is vitally important.

How to buy and sell Bitcoin?

There are various ways to buy and sell Bitcoins, such as online, ATM machines, direct contact with sellers, Bitcoin events and meetups.

For example, if you want to buy directly from the seller, you need to exchange your address or public key with the seller to transfer Bitcoins. There are websites that connect buyers and sellers like www.LocalBitCoins.com. This may not be the most efficient way, however, and you may end up paying more. The other option, which is more efficient, is buying it from an exchange. Exchange is basically a virtual meeting place for people to buy and sell at an agreed market price. This is similar to how the stock market and the futures markets work. There's an exchange where you have a number of people looking to buy and a number of people looking to sell. Through that you have what is called price discovery, where you can find out what the true market value is, based on supply and demand.

How do exchanges work?

Basically, what an exchange is and what they do is to provide an efficient environment for people to buy and sell cryptocurrencies. This isn't just for Bitcoin. You can also exchange Litecoin or other alternate coins. The website or the exchange takes a percentage of the trade, anything usually from a tenth of a percent to as high as one percent. So, it can be expensive if you're not careful. You really need to make sure that you're trading through the right exchanges.

To get this started, you sign up with an exchange to start trading. Some of them do require verification, especially if you're going to link it to a bank account or start moving bitcoins into fiat currencies.

Let's talk about some of the popular exchanges. Please note that these websites may change over the next few years, as Bitcoin is a new environment and there's still a lot of room for improvement.

5 best tools to start trading Bitcoin

In this section I'm going to be talking about my favorite tools and resources for trading cryptos. Bitcoin and a lot of these cryptocurrencies are traded on multiple exchanges and actually are traded at slightly different prices depending on which exchange they're on. One of the most common questions that I get is that what is the best exchange to trade Bitcoin or alternative currencies?

1) First tool to start with

The first exchange is Coinbase (www.coinbase.com). This is a great way to get started buying Bitcoin for beginners. Coinbase is one of the biggest cryptocurrency exchanges. Currently, they allow you to trade a few main cryptos including Bitcoin, Ethereum, Litecoin, Bitcoin cash and more. These coins are some of the largest market cap cryptocurrencies. So, you have more than enough access to the cryptocurrency market simply through it.

The good thing about Coinbase is that you can link this directly to your bank account and buy and sell easily. There's a US dollar wallet that you can actually send money into this account and it will sit there as cash. It doesn't have to be sitting invested in Bitcoin or other coins. It can just be sitting there in US dollars waiting for you to see a good opportunity, which is nice because obviously the US dollar is much more stable than Bitcoin.

2) Commonly used exchanges

Coinbase is not the best place for active trading, as the fees are pretty high, and they're not really set up that way.

If you want to expand your horizon on other types of cryptocurrencies, there are better exchanges, like Coinbase pro (previously called GDAX). Other examples of exchanges are Bitstamp, itBit, BTC-e, OKCoin, Bitfinex, Binance. These exchanges provide you access to trade a much larger selection of cryptocurrencies than Coinbase pro. The exchanges are different based on the country they operate in. The question you need to answer first is what country you live in, because certain exchanges have restrictions and availabilities in different places.

Just keep in mind that trading some of the altcoins can be very risky, as they can sometimes be incredibly volatile. Then it just doesn't even make it worth trading them. Another thing about altcoins is that you cannot buy them with US dollars, the only way to buy them is with Bitcoin. What that requires you to do is to first buy Bitcoin before you can purchase these. That means if bitcoin is losing value while you're holding it and trading altcoins with it, you will lose money when you're converting your Bitcoin back into US dollars. The fact that you're actually having to take two investments at the same time, to me, makes it unnecessarily complicated. However during the bubble it made it in some ways exponentially more profitable because both were going up at the same time, so it is one of those things that helped amplify profits for traders in the second half of 2017.

Because of that risk and complication of the process, it's usually a lot easier to just use a website like Coinbase pro and just trade directly into the four major cryptocurrencies using US dollars.

In some exchanges like Bitfinex, you can actually short through, which means you can make profit if the price of Bitcoin goes down. This is called shorting, where you sell first and then buy back at the lower price.

3) Finding the best cryptos to trade

You can trade cryptos 24 hours a day, which is a big plus. Throughout the day, they have lows and peaks of volume; they do correspond a little with the opening of the US market. From 9:30 to 10:30 or 11:00 Eastern Time we see an increase in volume as the US market starts trading. We see similar increases as other markets open because a lot of traders - whether you're trading US market, the Asian market or the European market - trade most actively in the first hour. So within the first hour of a big market opening, we see a bit of an increase in cryptocurrency volume.

To find the best coin to trade, we need to find the most frequently traded ones, or in other words, which ones have the highest volume. Another factor is the change rate. We want to pick cryptos that have big up and down movements, not side to sides. In the stock market, we find those stocks in gap scanners. In cryptos space, what I usually like to do is start on this website www.coinmarketcap.com.

If you also want to take a look at what the most active exchanges are right now, this website is where you want to start. Just keep in mind, the exchanges listed there are always changing; there are new exchanges coming out, old ones close down and in general this is a space that's constantly evolving.

This site has a lot of information like the market cap, the 24-hour trading volume, the current supply, how many coins are currently out there and what the max supply is. If you click on Bitcoin and go to its page, there's a section called market. This will show you all the different exchanges or marketplaces that you can trade Bitcoin. What you want to do is sort by pair and find the currency that you use. If you are in the US, for example, it would be BTC/USD, as you trade in US dollars. If you're in Europe, it's BTC/EUR for Euro. Here as shown in the picture, you'll notice one of the most active exchanges right now is BitFinex.

#	Source	Pair	▲	Volume (24h)	Price	Volume (%)	Category	Fee Type	Updated
8	Bitfinex	BTC/USD		$119,326,350	$6,326.60	1.91%	Spot	Percentage	Recently
20	Bimex	BTC/USD		$59,211,928	$8,338.36	0.47%	Spot	Percentage	19 hours ago
21	Coinsuper	BTC/USD		$53,004,606	$6,322.35	0.46%	Spot	Percentage	Recently
24	Coinbase Pro	BTC/USD		$48,043,736	$6,308.34	0.41%	Spot	Percentage	Recently
29	Bitstamp	BTC/USD		$41,384,011	$6,323.60	0.36%	Spot	Percentage	Recently
45	itBit	BTC/USD		$24,651,671	$6,329.00	0.21%	Spot	Percentage	Recently
46	Kraken	BTC/USD		$24,313,473	$6,323.60	0.20%	Spot	Percentage	Recently
69	CoinsBank	BTC/USD		$15,225,160	$6,285.56	0.13%	Spot	Percentage	Recently
70	Cryptonex	BTC/USD		$14,981,852	$6,161.84	0.13%	Spot	Percentage	Recently
71	Nanex	BTC/USD		$14,813,951	$6,430.0s	0.13%	Spot	No Fees	Recently
86	Gemini	BTC/USD		$8,920,770	$6,311.39	0.08%	Spot	Percentage	Recently

Bitcoin markets

You can see under that we have Simex, Bitstamp, Coinbase pro and many more. Between them, I look to only trade with maybe the top five or half a dozen exchanges by volume. You don't want to trade in exchanges that have very low volume, because when the liquidity is low you'll get a lot of slippage in your orders. So you just want to go where the biggest volume is.

If you're just getting started, what I would suggest is go here on this site, sort by the highest market cap cryptocurrencies and look at what the biggest exchanges are. Then just get familiar with those exchanges, sign up get accounts, and learn what those exchanges do, if you can or can't trade with them and maybe just start trading with very small sized amounts.

Do not use and consider exchanges like bank accounts; these are not FDIC insured trading accounts like you would have at each trade exchange. These are still risky marketplaces, so you want to make sure that you keep a lot of your value actually in hardware storage such as Trezor or Ledger. Whenever you have your cryptocurrency on an exchange, you don't actually own that cryptocurrency. What you have is credits for that cryptocurrency. You don't own a cryptocurrency unless you have control over the private key.

4) Charting tools

After you get your account set up with the best exchanges for you, then you want to make sure that you have access to the best charting tools. If you want to be an active trader, then you need to use the best charting and order execution tools. The one I recommend is Coinigy (www.coinigy.com); this is an all-in-one trading platform that has charting and order execution for all of the important digital currencies and exchanges. You can link your accounts to Coinigy, actually place orders and track your trades in a single location.

It aggregates all of the currency pair and all the exchanges in one platform. It has listed all of the different exchanges, and when you click on the exchange you want to look at, it'll give you a list of all of the currency pairs that this exchange offers.

It also has a really nice user interface and charting tools, where you can add different indicators and strategies.

There are other charting tools like Bitcoin charts (www.bitcoincharts.com) where you can trade dollars vs bitcoin. For each coin, there's a current price, the chart, average, volume and 24hr average price.

5) Join a trading community

Most people that attempt to trade lose money. There are many reasons why such a large percentage of people do so, but the top reason is a lack of training and education. Trading is a performance activity like golf, poker or other things. If you don't take the time to learn and practice the most effective strategies and techniques, then odds are you will lose money to more skilled traders. You need a solid strategy to pull the trigger on an entry and how to effectively trade.

Bitcoin can be traded 24/7. It's very helpful to set up your mobile charts and alerts. You can have price alerts in place by phone apps like Zero block or Coinbase, where you can frame prices, giving you an advantage if you intend to buy and sell at specific prices.

Four trading rules

Successful trading and investing is all about picking the right market direction at the right time.

Here I want to give you some of my top rules if you're new to cryptocurrencies or even if you have been in this space for a while. It can be extremely profitable if you do it right, but you can also lose a lot of money if you do it wrong.

Rule #1: plan your trades before you take a trade. Making money in any market is all about anticipation and understanding mass human psychology. There's an old cliché by Warren Buffett that says, 'be fearful when others are greedy and be greedy only when others are fearful'. So, don't chase emotional buying. This is why so many people lose money in almost every market. The scary moment is when people chase emotional rises in price. Nobody wants to buy in a market when things are cheap, but when it's hot, it grabs everybody's attention. That's the reason why trends go much further than most people anticipate. So, make sure you fully plan your trades.

Rule #2: avoid over-trading. This is probably the biggest mistake that I see new traders and investors make. You don't need to trade every little movement in the market. In reality, you only need a couple of good trades in the period you're trading to make a career's worth of profit.

Rule #3: manage your trades with no regrets. It's not always possible to pick tops in a market, and I never recommend targeting for tops or even lowest if you're shorting. You never know when the market is going to rollover, or if you are in a short, when it's going to bounce. So, if you're buying Bitcoins or other altcoins and you're trying to find when to get out, you want to manage that trade in a way that no matter what happens to the market (if it crashes on you or keeps going straight up), you don't lose money. Because you can't control the outcome of a market. In simple language, when you plan your trades and make some profit, you want to close your trade and put the money in your pocket, because the market doesn't always move toward your favor. If you make good profit and you don't take something off the table and then it crashes, you will regret that.

Rule #4: control the risks. Stay away from the buy and hope mentality. My thesis is to go to the market with some knowledge. When I trade or invest on something that I think there's a high probability of success, I can control my risks. Right now, there's a ton of hype in the Bitcoin market. Approach the market with knowledge and try to anticipate the hype and when the market is going to panic.

You also need to consider the risks of Bitcoin and other cryptocurrencies, as they are still in their experimental phase. My main recommendation to you is do not trade Bitcoin or invest in it with money you can't afford to lose.

Join a Bitcoin or trading community. Trading and investing can be a lonely place. It's helpful to be around other like-minded people who can support you or you can learn from.

CHAPTER 6:

How to store and use Bitcoin; Bitcoin wallet

If you have made the decision to buy some Bitcoins, you may now be asking yourself how to store them. There are bitcoin wallets to store your bitcoins, which I'll explain how they work, as well as different types of wallets and how to choose the best one. A bitcoin wallet is a software program where Bitcoins are stored. Wallets are also used to send and receive bitcoins or monitor bitcoin balances. Just like you need an email program like Outlook or Gmail to send and receive emails, you need a bitcoin wallet to manage your Bitcoin. Wallets interface with Bitcoin blockchain. They monitor Bitcoin addresses on the blockchain and update their own balances with each transaction.

To be technically accurate, Bitcoins are not stored anywhere; there is a private key or digital signature for every Bitcoin address that is saved in the Bitcoin wallet of the person who owns the balance. What defines a wallet is where its private key is stored. It's from this key that your wallet gets the power to send bitcoins to other people. Whoever knows your private key can have access your Bitcoins.

Digital signature or private key is supposed to be like a regular signature on paper, but in digital form. The main characteristics of a signature is that you can make it, but anyone who sees it can verify that it is yours and it is valid. The second thing is that the signature has to be tied to a particular document. We don't want our signatures being taken off from documents and being glued onto another one. Bitcoin uses a particular digital signature scheme called ECDSA (Elliptic Curve Digital Signature Algorithm). It relies on extremely hairy math. This algorithm has good randomness which is essential in digital signatures. If there's no good randomness, or in other words there's a pattern in generating the keys of your digital signature, then there's a big possibility that your signature could be hacked.

In practice, when we talk about how to store or spend your Bitcoin, you don't need to be worried about how it's stored in the blockchain and how the transaction happens. The main thing we have to keep track of is how to store secret signing keys. Security is the critical fact, because if someone can find or hack your signing keys, they get the power to steal your coins and spend them. The signing keys should also be kept somewhere that you can have access to whenever you want to spend your Bitcoin.

There are various types of wallets like web, mobile, desktop, hardware, paper, brain, multisig and HD wallets. I'll explain each one in a simple way, and help you choose the best one based on your need. In general, we can split wallets into two categories; hot and cold storages.

Hot and cold storages

A hot wallet refers to any type of Bitcoin wallet that is connected in some way to the internet. One convenient way is keeping your private keys on a local device, like keeping money in a physical wallet. The easy way of storing your keys is saving them in web services wallets or in a file in your local device such as your computer or your cellphone. This method is very convenient, as you can easily spend your coins by using an app in your cellphone. But security can be a concern, and your coins can be as secure as your device. If you lose the device, if it gets stolen or crashes, or if someone hacks into your device, you will lose all your coins.

The other method of keeping private keys is cold storage. In cold storage, your signing keys or coins are stored in an offline storage, or in other words it's not connected to the internet. Some example of cold storage are hardware wallets, paper wallets and brain wallets. It's safer and more reliable, but it's not convenient. The equivalent of cold storage is putting your money in a safe instead of carrying all the money in your wallet. We need separate keys and separate addresses for using hot and cold storage strategies. The question here is how to transfer coins back and forth between these storages. Each side needs to know what keys the other side is using. Besides that, each side needs to know the addresses at which the other side will accept transfers. Then the system lets you transfer back and forth. As I explained, cold storage is not online. The advantage of cold storage is that even if it's offline and not connected to any network, the hot storage still knows the addresses and can transfer coins into it whenever the amount of coins in the hot storage gets uncomfortably large. You don't need to risk and connect your cold storage to receive the coins at the time of transfer. Next time your cold storage is connected, it'll receive the information about all the transfers.

HD wallet

As Bitcoin wallets developed gradually, HD (Hierarchical Deterministic) wallets were created. HD wallets generate an initial phrase known as a seed up to 24 words. This seed is a string of common words that you can memorize, instead of the long confusing private key. You can enter the seed in order to reconstruct the private key. An HD wallet can also create various Bitcoin addresses from the same seed. All the transactions sent to addresses that are created by the same seed will be part of the same wallet. Because these private keys and seeds have complete power over your Bitcoins, they must be kept secret and safe. If you fail to protect your wallet's private key or seed, your Bitcoins can be lost forever. A standard Bitcoin wallet will create a wallet.dat file containing its private key. This copy has to be backed up by copying it to a safe location like an encrypted drive on your own computer, an external flash drive or even copying it on a piece of paper and hiding it somewhere.

Web wallet

As I mentioned earlier, hot wallets are the most popular but the least secure, since they allow access through internet connections. Markets, exchanges, betting sites and other Bitcoin services frequently require you deposit funds into their online wallets to be able to conduct your business. These web wallets are the least secure options for storing Bitcoins, since you don't have any access to your private keys. You're basically asking someone else to hold the coins for you. These wallets are also more vulnerable to hackers, as they have many loop holes along the way.

For example, the website, the device you're using to connect to the website or the internet connection can be monitored to steal your Bitcoins. This forces you to rely upon both site operators' honesty, and their security practices. In the event of internal fraud or external hacking, your bitcoins will be irretrievably lost. On the other hand, web wallets are highly convenient, as they allow you to buy, sell and send the bitcoins easily. For storing any significant amount of coins, web wallets are not worth the risk. My recommendation to you is to only keep a small amount of coins in web wallets as you would keep only a small amount of ready cash in the wallet you carry on you.

Desktop wallet

Another type of wallet to consider is a desktop wallet. These types of hot wallets store your private key on your computer. So, as long as your computer is free of malware or any security weaknesses, your Bitcoins are safe. But this is not the case for most of us. Desktops that are connected to the internet are easy and valuable targets for hackers.

Mobile wallets

These wallets store your private key on your mobile phone. Many wallets are accessible via mobile apps. Mobile wallets are highly convenient, but it is the worst possible scenario for security as they offer low protection and terrible privacy. There's a possibility to associate between your bitcoin wallet, your phone number and geo location. Plus, there is a high probability that phones get stolen, lost or broken. So, my advice is to password protect your wallet and create a private key backup. Substantial coins should not be stored in your mobile wallet.

Paper wallet

In this method the information gets printed on paper and can be stored in a safe. By keeping your private key on a piece of paper, only someone who can physically access that paper can steal your Bitcoins. However, paper wallets can be easily destroyed. Therefore, it's advisable to create multiple copies. So that if one is lost your bitcoins can still be retrieved.

Hardware wallets

These are physical devices which safely store your private key. They cannot be hacked, even if your device is compromised with malware. You can even use them with a public computer that you don't trust. Most hardware wallets provide a seed backup in the event of the device being lost or stolen. To send your bitcoin to someone with a hardware wallet, you'll need to have it connected to a computer and use some sort of webpage that allows control over the wallet. Hardware wallets provide both security and ease of use. Their only limitation is that you need to keep your hardware wallet on you at all times in order to send the coins.

Brain wallet

In a brain wallet, we take the information we want to protect and encrypt it under a passphrase or password that a user remembers. Whenever the user gives the passphrase in the future, then the information will be decrypted. Brain wallets have a significant disadvantage; they have a higher probability of being hacked. This is because people are usually very predictable in what they use as passwords, and hackers have way of knowing that.

Multi-signature

Multi-sig is an important feature some Bitcoin wallets have. It stands for multi-signature. A wallet that allows the sending of Bitcoins only with the approval of enough private keys out of a set of predefined ones. Imagine 3 people want to open a business together and invest some of their Bitcoins, but none of them wants only one person to have the private keys to this money. They each get one key and use a multi-sig wallet that requires two out of three of those keys. This way, none of them can run away with money, but they also don't need all the members. Multi-sig doesn't have to be only two out of three; it can be almost any combination. For example, a couple wants to have a shared account and decide that only if both of them agree to a purchase then they can spend the money. Or a company's board of directors that allows payments only by vote of the majority.

Now you know all about different Bitcoin wallets. Let's see how to choose the best wallet for your needs.

Different people use different wallets for different purposes. For example, if I need to store a large number of bitcoins safely, I will use a different wallet than if I just want to have some small Bitcoin change to pay for a cup of coffee. You need to decide which wallet works for you. Some of the questions you should ask yourself include how many Bitcoins will I be storing? How frequently will I use the wallet? Can I afford to pay for a hardware wallet? Do I need to carry the wallet around with me? Do I need to share the wallet with someone else? How much do I value my privacy? Depending on your answers to these questions, it will make it easier to decide the wallet that will work best. You may want to use more than one wallet. You can use a hardware wallet for large amounts of Bitcoins and also have a mobile wallet with a small balance in it for daily payments. This way, even if your mobile phone breaks or gets stolen, you're not risking a lot of money.

CHAPTER 7:

Bitcoin mining and earning cryptocurrencies

Bitcoins are generated using very powerful computers solving very complicated mathematical equations. Bitcoin is a network of computers that maintain a blockchain. The bitcoin blockchain is the public ledger which is similar to the history book of bitcoin transactions. This ledger documents and saves all the bitcoin transactions that have been ever made. This history is vitally important, as it keeps all the bitcoin accounts or addresses. The important question here is how we can trust this ledger. Remember, bitcoin doesn't have a central authority to approve and verify the bitcoin history book, which means that anyone can write it. So how can you have a reliable history book that anyone can write, and nobody approves?

Without a central authority to monitor and approve transactions, we could have many different versions of this book and we wouldn't know which one is correct. In order to make this book reliable, we need some kind of voting system. This way we can all agree on the right version of the history book and make sure that nobody can profit by causing confusion.

Traditional voting systems require a central authority to use identification and determines who has the right to vote. This can't be the case with bitcoin. As part of decentralization, bitcoin mining is open to everyone. So, this voting system needs to allow anyone to take part, but at the same time prevent fraud. To do this, bitcoin mining is used. The bitcoin protocols grant new bitcoins every time someone adds a new transaction to the history book. It's also known as adding a block to the blockchain. It requires a very high capital cost and computational power. Bitcoin depends on miners to store and broadcast the blockchain and to validate new transactions. What do I mean by computational power and why is it needed? Who are miners?

There's a mathematical way to prove that you took data, like a page in a history book, and ran some calculations on it. This approval process requires computational power and electricity that costs money. Anyone with access to the internet and suitable hardware can participate in mining. So, if you have a computer at home and are connected to the internet, you can start mining today.

The process works this way, if everyone agrees to add your page to the history book, then you'll get paid for your work. But if they don't agree to add the page, then you just wasted money writing it. This ensures that everyone writing this history book try very hard to agree with each other. The high cost of writing up fake versions keeps fraud away from the system.

This may sound like a weird way to solve the problem. But really it makes a lot of sense when there's no central authority. Miners have to invest money in mining as a kind of collateral. They get their money back in bitcoins if their mining was good for the network. Because miners are paid in newly issued bitcoins, it's sort of a cost that all bitcoin holders pay in the form of inflation, in order to secure the network and to make sure that the history book is accurate.

This model solves two problems; it decentralizes the trust in the ledger so that we don't need a central authority. It also lets the protocol fairly create and distribute all the bitcoins that need to exist instead of them all going to the inventor. One of the core advantages of Bitcoin is how the coins are created. Once again, they're not created all at once by the inventor as that would be unfair. Rather they're created slowly over time and paid to all those who take part in processing blocks of transactions and adding them to the blockchain.

The rate and limit of coins created are clearly defined in advance; 50 coins per block of transactions that is added to the blockchain, which takes roughly 10 minutes. This number decreases by half each two hundred and ten thousand blocks, or roughly four years, for a total limit of 21 million coins. Importantly, the calculations that miners do are self-adjusting to maintain this rate of one block per 10 minutes.

Every time you use a Bitcoin wallet, you are reading through the entire blockchain and verifying all the information including the cryptography computational power and very importantly the creation of new bitcoins. So, reading the history book keeps miners warned and if anything is out of place or illegitimate you toss it out and find a version of the blockchain that does follow the rules.

Making money from mining

A miner is a computer of some sort that runs the mining algorithm, but the person who owns the computer is also called a miner. When you hear the word miner it may mean the computer or the person or both, depending on the context. So, what does a miner do? Miners take the transactions that haven't been processed yet and put them together in a block. That's the new page we're adding to the history book. Remember we said that it requires computational power. The miner needs to run some calculations on this block and these calculations have a difficulty that takes time to complete. This difficulty is that blocks are completed once every 10 minutes on average, but it's quite random. So, what happens is a kind of lottery where all the miners are trying to complete these calculations until one of them succeeds. The one that succeeds gets the reward of new coins granted from the protocol every time a block of transactions is completed and added to the chain. Then, all the miners start working on the next block. Of course, the faster your computer is the more likely you are to solve the calculations first. But remember you only get a finite number of bitcoins per block. So, based on this method, it isn't worth investing in computing power if you only get randomly payed.

To summarize, the Bitcoin protocol is paying people to invest computational power in the blockchain but promises no profit. Back in the day when Bitcoin was practically unheard of, there were only a few miners. These people could find a block once every 10 minutes on average and so they used their laptops and desktops to mine for bitcoins. The difficulty was low, and nobody did this competitively because bitcoins were worthless back then. However, as the exchange rate of Bitcoin started to rise, other people realized that they could mine bitcoins and sell them for profit. More miners joined the party, and this made it harder to find blocks.

Faster computers would generate more bitcoins than slower computers. Instead of using regular computer processors, it's better to use graphic card processors and FPGAs to get more powerful mining capabilities. This may not be the best or most efficient way, as the difficulty continues to grow. A more efficient method would be ASIC mining. ASIC stands for application-specific integrated circuit. These are computers that are designed specifically to mine bitcoins and they're much better at it. Today, profitable Bitcoin mining is done only with ASIC miners. Remember the protocol will still generate bitcoins at the same rate. So faster computers will get a bigger share of the cake and slower computers will get a smaller share of it, but the size of the cake won't change.

Mining pools

As the Bitcoin network gained more mining power, the difficulty grew and the average time of ten minutes between finding blocks remained. Every block reward was now worth more and there were more people competing. So as a miner you'd find yourself finding fewer and fewer blocks but each block was worth more and more. Instead of a steady business where they would get paid a little once a day, miners were playing a sort of lottery where they would get paid a lot once they got lucky enough to find a block. But business people prefer a steady income over lottery tickets and so they came up with an idea of mining pools. A group of miners joined together to find blocks and when one of them found a block, they would split the reward with everyone in the pool. They effectively get the same profit, but it is much steadier and less random at the same time.

Cloud mining

Another way of participating in mining is through cloud mining. This means that instead of buying a miner and running it yourself, you'd lease mining power from companies who claim to own large mining farms and want to diversify their income streams. While this can in theory be cheaper and more efficient, many of these services are quite shady and it's very difficult to verify that you're getting the service you're paying for. If you're not an expert, a good rule of thumb is to be skeptical of these services.

Bitcoin mining is a very technical field in a very competitive industry. The inventor of Bitcoin managed to design a system that gives everyone a fair chance to take part in it every step of the way and that is what is so innovative about Bitcoin.

Mining can be a great way to start getting involved in cryptocurrencies, especially if you already have a computer system at home. You just need to have the right graphics cards and do some minor adjustments. Then just let the computer run and make money for you. In the next section, I'll explain how to start mining and set up your fist mining rig.

How to start mining

As I described earlier, during mining, you take your computer and run a program on it; the program uses your computer horsepower to solve the math. Based on how much data you process, you get paid in cryptocurrency in return for your service. You give the cloud your CPU or GPU and they give you currency, which you can sell the currency for cash. This provides a smart way to issue the currency and also creates an incentive for more people to mine. Since miners are required to approve Bitcoin transactions, more miners means a more secure network. The Bitcoin network automatically changes the difficulty of the math problems as I explained earlier in this book.

If we solve one of the puzzles, we can unlock a Bitcoin block which at a few thousands per coin, obviously there's a huge incentive there. What that's created is a surge in Bitcoin mining operations to the extent that the power use of the Bitcoin network is about 30TWh a year. That's more energy than Ireland used last year. Despite the fact that Bitcoin price dropped sharply in 2018, the hash rate - essentially the rate of mining going on - has increased drastically by 155% since the start of the year. The main reason for the increase in hash rate is that each block rewards of 12.5 bitcoin. However, this will not last forever. On May 25, 2020, the next so-called halvening will occur, decreasing block rewards down to 6.25 bitcoin per block. So there's a rush to mine as much bitcoin as possible. Increase in hash rate evidently shows that faith in bitcoin hasn't been lost yet.

Right now, Bitcoin mining is mainly done in huge Bitcoin farm operations. Most of these farms and warehouses are in China or Russia.

Mining hardware

Let's first look into the main mining hardware modules that miners are using today.

CPU:

In the early days, Bitcoin miners solved the mentioned math problems with processors and their computers. CPU was actually the only way to mine Bitcoin initially, and it was quite effective as the complexity of solving the math was low. Any old laptop could mine hundreds of Bitcoin a week (worth millions of dollars at today's value). It is still possible to mine some altcoins with CPU. One of these altcoins is Monero (XMR).

GPU:

As complexity of mining raised, soon miners discovered that graphics cards used for gaming were much better suited to this kind of working. Graphics cards are faster; they handle very complex mathematical calculations to run videos in your computer. The drawback is that GPU uses more electricity and generates a lot of heat. It could be actually useful if you want to heat up your room while you're earning money, which is what I do right now. Ethereum can be mined with GPU. The best and most efficient GPU for this purpose is AMD. I'll explain later how to set up your mining rig.

ASIC:

ASIC (Application Specific Integrated Circuit) chips are designed specifically for Bitcoin mining. They can't run any other tasks than mining Bitcoin. Initially, dedicated mining systems were FPGAs (Field-Programmable Gate Array), but soon they were replaced by ASICs, as they were more powerful.

ASIC technology has made Bitcoin mining even faster while using less power. These devices are unbelievably powerful and efficient. This is one of the reasons that regular miners can't compete with them with regular computers at home.

ASIC system may not be the best option for beginners, as the cost of each unit is quite high, and it makes a lot of noise while it's running. Also, as it runs only one task, it could be a little hard to sell it if you ever need to. The number of Bitcoin ASIC hardware manufacturers are limited today; BitFury Group and Bitmain Technologies Ltd. are the two major companies.

If you think Bitcoin mining is getting too competitive and also pricey to start, don't worry, you haven't missed the boat yet. If you just want to get involved in it before it becomes too complicated again, start mining Ethereum by using your computer at home.

Setting up a mining rig

Now, it's time to explain how to actually set up your system and run it. There are various ways to set up a miner, but I'm going to go over a popular way to mine Ethereum. It can simply be set up with regular computers at home.

Mining Ethereum:

This guide will help you to set up your own GPU for mining Ethereum. First, I'll list what hardware you need to set up your system.

GPU:

GPU is the main part of Ethereum mining rigs as described earlier. The best GPUs for this system are Radeon RX 4xx/5xx series or Nvidia GTX 1060/1070.

Motherboard:

What motherboard to get? For building 6 GPU mining rigs you could get Z170 and Z270 Pro series. Other cheaper and reliable motherboards like H81 pro BTC 2.0 work well too. It depends how much budget you want to spend.

For the 12/13 GPU mining rigs, the best motherboard to use is an ASRock H110 Pro BTC. Windows can only support 12 AMD GPUs at the same time. You can also add 6 Nvidia GPUs to your 12 AMD system and run 18 GPU. Based on my experience, it's not really a good idea. There are some Motherboards that support 19 GPUs, but based on personal experience, you will face some problems when your system is operating. So it's better to stick to 6 or 12 GPU Mining Rigs.

RAM:

You can run your mining system with a simple and inexpensive RAM (1333MHz). 4GB of any RAM is enough for 6 GPU rigs or 8GB of any RAM is good for 12/13 GPU rigs.

CPU:

Processor is not playing a big role in mining either; the main power for solving the math is taken from GPU. Therefore, any CPU that fits the socket of your motherboard works fine. For example, Intel processor 1150 for H81 pro BTC, or 1151 LGA CPU for H110 Pro BTC Motherboard is sufficient. For 12/13 GPU rigs, you need to get a slightly better and more expensive CPU (60-100).

PSU:

PSU is one of the main components of your mining rig that provides power to the system. A cheap and low-quality PSU can cause serious damages, especially if you run your mining rig 24/7 at maximum power. From my own experience, I have heard many times that a cheap PSU has caught on fire or even melted. If you already have expensive components in your system, the worst thing would be saving money on your PSU. So, I recommend spending more money here to get high quality and reliable PSUs; only 80 Plus Gold or 80 Plus Platinum certified PSUs.

Make sure PSUs are not overloaded. For example, a 1000W gold PSU should never be forced over 800W power draw, you may end up destroying it.

80 Plus certificate tells you how much extra power a PSU draws to power your components. Certified products have more than 80% energy efficiency. In another word, the waste is only 20% or less, which reduces electricity use and bills compared to less efficient PSUs. If you have already invested so much on your mining

rig, there is no point saving a small amount of money on a very important component. By using a high-quality PSU, you will save power in the long-term that will increase your ROI. Let's say you have a system with 8 GPUs that uses exact 1000W, 80 Plus Platinum will use 1060W total (the waste is 60W), 80 Plus Gold will use 1080W, 80 Plus Silver will use 1100W, 80 Plus Bronze will use 1120W and a PSU without a certificate will use 1150W or more. The power draw from each hardware module is the same, the only thing that is changing is the extra power the PSU needs to power your rig. In our example, our system always uses 1000W, but the PSU needs extra to generate the required power. Higher grade certificates are expensive and most of them have 5-10 years warranty. It's very important to get a top grade and high-quality PSU to minimize the risk.

The last thing about PSU, Never connect more than 2 GPU risers to the same PSU Molex/Sata cable. This is very important because it could melt the cable or burn the PSU.

Risers:

Risers are expansion cards that are used to extend the slots on motherboards, as there's not enough space on motherboards to mount all 6 or 12 GPUs on them. Risers could be the reason for most of the problems with your GPUs. For example, if you have a problem with one of your GPUs, and you can't see it or can't flash the bios, try to replace the riser. It's better to get high-quality risers (latest release), as they are a very important part of your mining rig. Usually, 60cm cable length is fine.

Disk:

You don't need to spend so much on a disk. For a 6-GPU and a 12-GPU mining rig, a 4GB SSD and 128GB SSD should be enough.

OS (Operating System):

For the operating system, I recommend using Windows 10. Most people think Linux is a more stable operating system, or because it's so lightweight, it runs better. It would seem logical, but Linux has various problems:

- Linux isn't more stable than Windows 10; if Windows is set up properly, it can run for months without the need of a reboot, most people don't know that.
- Higher power draw. Under Linux your GPUs will use about 5% more power than on Windows, with the exact same settings for overclock/undervolt. Overclock/undervolt has to be done on GPUs to lower the power consumption and increase the RIO.
- Drivers for the GPUs under Linux are outdated and they have poor support
- Overclock tools for Linux are hard to get. They most likely won't work properly on all GPUs, especially not the undervolt part, which is the most important part.
- In Linux, it's hard to tell if one GPU is producing memory errors or if something is wrong with the GPU. This could cause to break the GPU in long term. Windows has strong tools such as HWinfo64 that can tell you precisely whether or not your GPU is working properly.

- As I described earlier, in mining, we are computing a mathematical problem or hash function. The speed at which a computer is completing an operation in the code is called Hash Rate. A higher hash rate is better in mining, as it increases your opportunity of finding the next block and receiving the reward. In this regard, Linux is not efficient and you will not get the same hash rate as on Windows. For example, the exact same settings on a GPU on Linux would give you 28.8 MH/s and on Windows, you will get 29.3 MH/s.

- You could optimize your Windows by updating Windows and then stopping it from automatic updates. You don't want the system gets updated randomly. Change the power plan option to "never turn off the display" and "never put the computer to sleep". For the sake of keeping this instruction short, I won't explain the details here. The instruction can be found on google.

These are all the hardware modules you need to start mining. On my website, I have listed what I am using in my mining rig, which is cost effective and very efficient. Check out my website for more details www.futurentech.com.

Windows Updates

If you have downloaded Windows 10 from the official Microsoft website then your Windows 10 pro is almost up to date.

- Go to the Control Center
- Go to Updates and Security

- Go to Windows Update

- Check for updates and install the latest ones

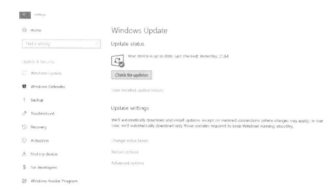

Drivers and Software

Download latest .NET Framework 3.5 Offline Installer – it is
required to run Polaris 1.6 and OverdriveNtool. Windows 10
Comes with 4.x Framework, but that one will not work with Polaris
and OverdriveNtool. You need to install the .NET 3.5 manually.
Insert the Windows 10 USB stick into the PC (the one you used to
install Windows) and set the USB Disc Drive as the Source for
installing the .NET 3.5. .NET Framework 3.5 Offline Installer is the
Guide how to do it, it's very simple.

Download latest Drivers for your motherboard, especially
latest **Chipset driver. This is very important**.

Windows 10 Registry Tweaks For Mining

- Once windows is fully installed and has booted for the first time, you will have to run the **Windows 10 Registry tweaks for mining.bat** file. Using this tweak disables everything that is not important for mining on Windows. **Disable everything from this tool.**

  - Run it in administrator mode

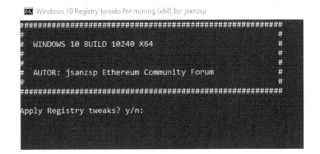

Windows Registry Tweaks

  - Once everything is applied, restart the rig

Stop Windows Update

- Open RUN and type in it "**services.msc**", the services page will be opened

  - Search for "**Windows Update**"

  - Click "**Stop**" if the status is "**Live**" or "**Checking/Running**"

  - Select "**Disabled**" on "**Startup type**"

  - Apply & Restart

  - Do it again and make sure that it is disabled

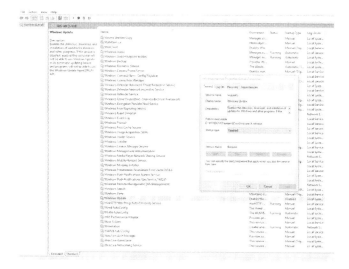

Services Tweaks

Increase Virtual Memory

- Open up search and type in "**This PC**", right click on it and choose "**Properties**"

- Then click on "**Advanced System Settings**"

- On "**Advanced**" tab where it says "**Performance**" click on "**Settings**"

- Click "**Advanced**" tab

- At the bottom you will see the "**Virtual memory**" option, click on "**Change**"

- Uncheck "**Automatically manage paging file size for all drives**"

- Check "**Custom size**"

- Now depending on how many GPUs you have you will need to increase the virtual memory. In general, for each GPU you need to have 3GB Virtual Memory (because the DAG file used for hashing Ethereum is above 2GB and going to 3 GB this year). So if you have 6 GPU you need to have 6x3GB Virtual Memory = 18GB

- In the "**Initial size (MB)**" type: **2000MB x GPU Count (12000MB for 6 GPU mining rig, 24000MB for 12 GPU)**

- In the "**Maximum size (MB)**" type: **3000MB x GPU Count (18000MB for 6 GPU mining rig, 36000MB for 12 GPU)**

- Click "**Set**" , then "**OK**" and "**Apply**"

- Restart the rig

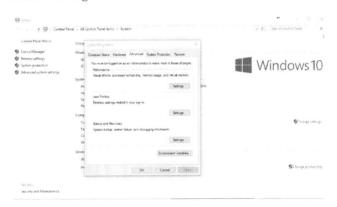

Windows Advanced Settings

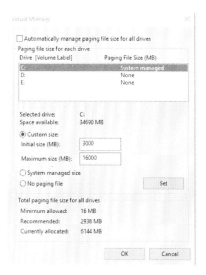

Changing Virtual Memory

Power Plan and Options

- Open up search and type in: "**Power Options**"

- Click on "**Show additional plans**"

- Select "**High Performance**" plan

- Then while still on the "**High Performance**" plan, click on "**Change plan settings**"

- Select "**Never**" on all four selections for "**Turn off the display**" and "**Put the computer to sleep**"

- Click "**Change advanced power settings**" and look for "**PCI Express**" -> "**Link State Power Management**", make sure it is set to "**OFF**", usually it is, but better to check on that

- Restart the rig

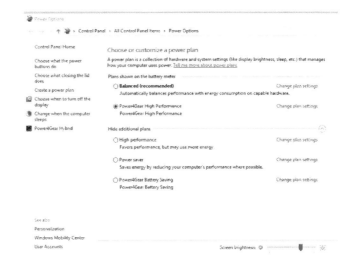

Change Power Options

Change Power Plan

GPU drivers

Now after Windows is setup properly, download a tool called DDU. That tool will uninstall your current driver (even your integrated GPU) and block Windows from automatically installing GPU drivers. That's important so that Windows does not install an outdated driver. It will ask you to run in "safe mode" but that is not necessary. When you run the program just click on "Clean and restart". We want to manually download and install the right drivers.

Block Windows Automatic Driver Installation

Now turn off your PC and connect only **ONE** GPU.

What driver is the best?

IMPORTANT – From AMD Crimson 17.10 driver and all drivers released after that, they have added the mining mode in the driver and enabled up to 12 AMD GPUs to be able to run on Windows 10.

AMD released Radeon Software Adrenaline Edition, download latest version of it. It will improve hashrate on some cards, and in general, will give you best possible hashrate on all RX 4xx and 5xx cards.

Sometimes, it's possible that you get better results with the Beta Blockchain Driver , but that driver only supports 8 AMD GPUs and please try first the Adrenaline edition. In the Blockchain driver you will not need to change GPUs to Compute mode, it's there as default.

If Adrenaline or Blockchain driver is not working, your last hope is latest Crimson ReLive Driver release (you need to change in Radeon Settings to compute mode for each GPU).

At the beginning of the install process go to **CUSTOM** instead of Express and ONLY select "AMD Display Driver" and "AMD Settings". During install **SKIP** the installing of ReLive, because we won't need it.

Optional fix for RX 470 and RX 570

After you have installed the driver, restart your PC. If you've already modified your GPUs before, there might be a possibility that you don't be able to see them anymore. That is most likely a problem with the RX 570 series and it's very rare with some RX 580 models. The problem comes from the BIOS mod, because it changes how the GPU work. You will need to Patch your drivers to make them work properly or the driver will just end up disabling or hiding GPUs (Error 43). This is only needed if you can't see your BIOS modified GPUs in the Windows Device Manager. Download the Pixel Clock Patcher.

Pixel Clock Patcher for RX470/570 GPU

Installing all other GPUs

Once you successfully installed the driver with just one GPU, shut down your PC and plug in all of the other GPUs. After that, when you turn the PC back on, it should automatically detect each of them and it will install the drivers for all of them. Just remember that it will take some time (about 5-10 minutes) for all of the GPUs to be detected properly. You can open up the Device manager, to see if all of the GPUs are listed there. Just turn the PC on and wait 5-10 minutes before doing anything, Windows will do its job.

Mining (Compute) Driver Mode

Now after you have all of your GPUs under the right driver, there is one more important step to make.

1. Open up Radeon Settings

Radeon Settings

2. Then click on "Gaming"

Radeon Settings Gaming

3. After that click on "Global Settings"

Global Radeon Settings

4.   Now all of your GPU settings are displayed. We will need to change each GPU to be in **MINING** (Compute) mode. By default, each GPU is in graphics mode, and to enable proper mining (high hashrate due to DAG fix) we will need to change each GPUs "**Global Graphics**" "**GPU Workload**" to **COMPUTE** from Graphics. Each time you change one of the GPUs to COMPUTE, it will ask you to restart the AMD Settings, which will take some time. It's also possible that even if you've changed it to Compute, the GPU still displays "Graphics" instead of it. To fix this, try to restart your PC and see if the value was stored properly after that.

Change GPU Workload to Compute

BIOS mod

Each GPU has its own BIOS, which specifies how it should work. There are four different memory types that you will encounter on your GPU: Hynix, Elpida, Micron, Samsung.

During mining Ethereum, you will only be using memory of the GPU, That means the higher quality of the memory is, the better hashrate you can get. While testing all of the memory types, I've found out that Samsung and Hynix are a little bit better than Elpida and Micron, but the difference is small.

Hynix and Samsung Graphic cards:

- all RX 580 8gb cards

- most of RX 570 8gb cards (there is a chance of getting micron memory)

Elpida Graphic cards:

- most of RX 580 4gb cards (there is a smaller possibility of getting micron and very rare hynix/samsung)

- most of RX 570 4gb cards (there is a smaller possibility of getting micron and very rare hynix/samsung)

Micron Graphic cards:

- they are the rarest memory type and you can found them mostly on RX 570 8gb cards (they can appear on every RX series but it's very rare)

What memory are my cards and how to export the BIOS?

Download a tool called GPU-Z.

This tool allows you to see what memory type your card has as you can see in this picture.

GPU-Z BIOS Exporting

in the rectangle selection you can see the "Memory Type". In this example, it's Elpida. If you bought your GPUs all at once, and they are the same card type, and if you see they all have the same Memory Type, that means that they all **CAN USE THE SAME BIOS**. **Exporting GPU BIOS** can be done with clicking on the circle as displayed in the picture above, under the "BIOS Version". Now you have your original BIOS exported, make a backup before going to the next step.

How to BIOS MOD yourself in one click?

What do you need to know about BIOS mod?

There are various guides and tutorials how to BIOS mod your GPU and they are all very confusing and risky to use, especially if you accidentally use a BIOS that is not made for your GPU. **In BIOS we need only to change Memory Timings.** There are many reasons to do it that way, because **GPUs don't behave identical even if they are the same GPU; they can give much different results.**

For example if you have Sapphire RX 570 4gb, Elpida memory cards, they are all exact the same and the results you get from them:

- some GPU can handle only 1950 MHz on memory
- some GPU can handle 2125 MHz on memory
- some GPU can run at 800mV on memory voltage
- some GPU can run at 875mV on memory voltage

As you notice, there is a wide range of possible GPU settings, even if they are the same GPU. It would be very dangerous to put voltage values and memory clock values in the BIOS, because one of the GPUs could not handle it properly and you could even brick your card. It makes no sense to do it if you have software like OverdriveNtool which gives you full access to GPU and its voltages and clock rates. **There are no universal GPU settings**, please follow the overclock chapter in my Ethereum Mining Guide for the only right way to overclock and under-volt your GPUs. **Never change voltage and clock rates in BIOS.**

Anorak BIOSes have only memory timings replaced and the power save BIOSes have:

- voltage on memory reduced to 900mV because they can't know if your GPU can run at 850mV or even 800mV, the reason is explained above (for most BIOSes this has no effect, your GPU driver will restore it to default settings).

- GPU core clock rate reduced to 1100-1200 MHz

- TDP(W) reduced by about 20% (it has no effect if you use software for Overclock/Undervolt)

Those settings from anorak BIOSes are trying to be as general as possible and they are way over the actual limit of your GPU. The GPU core voltage should never be touched (in BIOS), people like to change the dynamic voltage to static ones (direct mV values) and that can cause GPU to get bricked if your PC freezes or you get random power outage.

Memory Timings?

Memory timings are the only part in the GPU BIOS that you need to change. Your miner software is using the memory of the GPU to make the calculations (hashrate you see in Claymore). In the GPU BIOS it is described how the GPU Memory should behave on specific clock rates. By changing how your memory behaves at higher clock rates (increase tick rate of timings) we can make the GPU calculate its operations faster.

By changing memory timings, we can increase the hashrate, depending how fast the new timings are. There is nothing else you need to change.

How to BIOS mod properly with one click?

Now after you know the basics of BIOS mod, you really don't need to know anything else because we will **only replace the Memory Timings in the GPU BIOS.** Overclock and undervolt can be done through software and we **NEVER** change those values in the BIOS, because it can brick your cards if you do. It's as simple as that.

Download Polaris 1.6.5. This software recently got a database of almost all memory timings and allows you to change your GPU Memory timings in one click. It auto detects and replaces them.

BIOS mod steps:

- Export your original BIOS with GPU-Z as explained in chapter "BIOS MOD".

- Open Polaris 1.6.5.

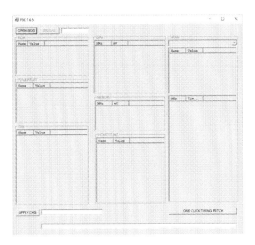

Polaris BIOS Editor 1.6

- Open your original exported BIOS from GPU-Z

Original BIOS

- Click on: **ONE CLICK TIMING PATCH**

One Click BIOS Mod

- Click Ok (depending on your memory type it will patch 1 or 2 different memory timings)

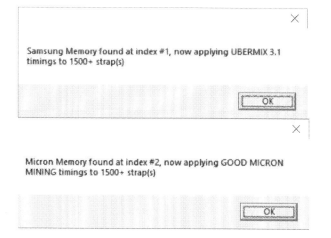

Samsung Memory found at index #1, now applying UBERMIX 3.1 timings to 1500+ strap(s)

OK

Micron Memory found at index #2, now applying GOOD MICRON MINING timings to 1500+ strap(s)

OK

Upgraded Memory Timings

- Save your new BIOS

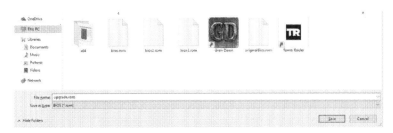

Save New BIOS

That's it, you have proper BIOS mod without making any unnecessary risks to your GPU. With OverdriveNtool you can adjust properly voltages and clock rates of your GPU individually to get maximum hahsrate for minimum power draw without risking stability or GPU's lifetime.

IMPORTANT – Always work with the original BIOS of your cards, don't download random BIOS online cause you can't be sure they are made for your card type. Even if they are the same model, it does not mean they have the same BIOS. It's very important to work with the original card BIOS to reduce the unnecessary risk to the minimum.

How to flash the new BIOS?

First you will need to download a tool for flashing the BIOS called ATIFlash.

With this tool you can put the custom BIOS over your current one. Always **make a backup for your current BIOS and store it somewhere safe, you can never know when you are going to need it.**

IMPORTANT – be careful what BIOS you are going to flash on what GPU. I would recommend you to never have different card types plugged in when you are going to flash, so you don't flash, by accident, a wrong BIOS to a wrong card. However, it is almost impossible, because if you use the AtiFlash properly as explained in this guide, it should give you a warning that you can't flash the specific BIOS, because it's a different type than your original card.

1.   Put the custom BIOS that you are going to flash in the AtiFlash folder, for example "upgrade.rom".

Upgraded BIOS, Part 1

2. Copy the file path location to the AtiFlash folder like shown in the picture

Upgraded BIOS, Part 2

3. Press Windows Key + S, then type CMD, after that the first result will be "Command Prompt", right click on it and press "Run as Administrator".

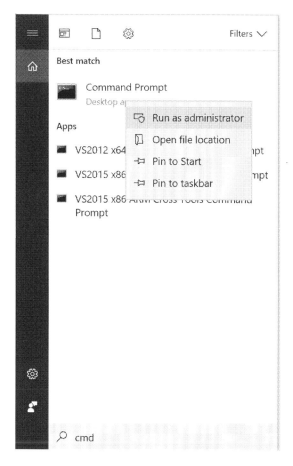

Run CMD as Administrator

4. Now in the console window, type "cd " and copy the path to the AtiFlash directory as you did in "Step 2", and it should look like in the picture (except your PC username is different than mine of course).

Change Directory to AtiWinFlash

5. Now type in the console "AtiFlash -f -p 0 upgrade.rom" where each word represents the following:

- AtiFlash -> the name of the execution file.

- -f -> forces the flash of a GPU.

- -p 0 -> the GPU at position 0 will be flashed (GPU orders are as displayed in GPU-Z or Device Manager).

- upgrade.rom -> that is the name of the BIOS that you want to flash to the GPU

Command to Flash GPU BIOS

BIOS flash problems

- you may get the error that the IDs mismatched; that means you want to flash the wrong BIOS on your GPU

- you may get the error that the ROM file could not be read; for that you will need to replace the Riser your GPU is connected to, or connect that GPU directly to the motherboard or use a DDU to reinstall the current driver and install the driver back on. If it does not work, try to download Atiflash 2.74 Version and try using this instead.

- you may get the error on 4000 or 8000 bytes wrong size; that means you want to flash the wrong BIOS. Some older cards have a 512kb BIOS; exact same new version of cards use a 256kb BIOS, which means you wanted to flash the older BIOS to the new type and of course you got rejected. You need to always be careful about what BIOS you are flashing.

- after flashing your GPU and restarting the PC, if you don't see the GPU anymore, means you didn't install the "Pixel Patcher". Please go to the GPU Driver section and read it again.

GPU Miner

Now after all your GPUs are flashed with the right upgraded BIOS, we can move on to the most important step, the mining software part. There are a couple of different common mining programs, depending on the algorithm they are working with, the most common ones are:

- **Ethash** (Ethereum, Ubiq, Ethereum classic) can be mined with the same mining software, because they all work under the same algorithm (ethash) and the best miner is Claymore 10.2

- **CryptoNight** (Monero, Electroneum) are also mined with the same software

- **EquiHash** (Zcash, Zclassic, Bitcoin Gold) are also mined with the same software

This guide is focused on the **Ethash** algorithm mining, so the settings and the tutorials from this guide are not optimized for the other mining algorithms like **CryptoNight.**

Claymore 10.2

Claymore 10.2 is currently the best miner for Ethereum, and it comes with a nice option of dual mining with some other altcoins (Decred, Sia...). That can boost your profit by around 20-30% for 20% more power draw. Even if you have expensive electricity, the bonus profit is probably worth it.

Is Claymore's developer fee a right thing?

Claymore software has a fixed fee of 1% when you are mining Ethereum or 2% fee when you are mining Decred. There are various problems that can happen due to the way the Fee is working. The fee works in a way that each hour you will be disconnected from your mining process and for about 1-2 minutes, you will mine for the Claymore developers. After that it will connect you again to your pool and start the mining again. By constant disconnecting and reconnecting each hour, your GPU cools down and then heats up again, and by doing that you are risking the life of your GPUs. I heard from many people that after some time one of the GPUs would reset to the default clock settings because of the constant disconnecting/reconnecting. It would also "hang" and crash the miner or cause it to recreate the DAG file, and you end up losing valuable time with that. Claymore is a really cool software and I think there could be a better way to support the developers, rather than risking our own miner stability. By using the official Claymore, I lost about 3% of my shares compared to using the Claymore without the Developer Fee, everyone can try it for themselves and see the difference.

Claymore 10.2 without Developer Fee?

On the bitcointalk forum, there are always releases for Claymore with the developer fee removed. Recently there was a 10.2 version released and confirmed to work (Claymore developers wrote that if you use a "hack" that would remove the developer fee from the program, it would autodetect it and give you a much lower hashrate as a punishment, but that was confirmed to be false).

You can download the Claymore with the developer fee removed: Claymore 10.2 NoFee Version

How to setup Claymore?

Claymore runs through its "start.bat" file. In the "start.bat" (you can open it with the notepad) you just need to write the following (no setx commands before that):

EthDcrMiner64.exe -epool yourMiningPool -ewal yourEthAddress -epsw x -dcri 6

Ethereum Mining Pool

-epool is the mining pool you can use, it's just a personal preference. Some people like to use nanopool, some like dwarfpool or ethermine. You can use any pool you like. Be careful what pool you are using; it should be based on your location. It would make no sense to mine on a European pool if you are in America. Always use the pool that is close to you. Nanopool, dwarfpool, ethermine and others have mostly location specific pools, you can't miss them; they mostly start with EU, US or Asia. After that you can write your own Ethereum address which is used to collect your Ethereum shares. You can view statistics on the mining pool by searching it with your address, for example if you are using nanopool you can see your current active statistics.

with: **https://eth.nanopool.org/yourEthereumAddress**. For example using Nanopool:

EthDcrMiner64.exe -epool eth-eu1.nanopool.org:9999 -ewal yourEthAddress -epsw x -dcri 6

Do not add "SETX" commands at start, they are not needed.

I use nanopool to mine Ethereum, you can use ethermine or dwarfpool also. Ethermine gives you most reliable statistics. Read following chapters to see why I use nanopool.

Ethereum Wallet Address

-ewall is your Ethereum address. Be careful, because you will always need to write only an Ethereum wallet address, not a Bitcoin or any other addresses. Most easy way to create an Ethereum wallet and keep it safe is to use the exchange sites like Coinbase, Bitfinex or Bittrex. They offer you high security and you can use the Two Factor Authentication which makes it very secure. For big amounts I would recommend to use offline wallets like Trezor Bitcoin Wallet.

How to setup dual mining?

Ethereum is mined just by using the memory of your GPU, so the GPUs core is almost not affected by the Ethereum mining at all. This gives a possibility to utilize the GPU core for mining some other coins in the same time as you mine Ethereum without affecting its hashrate. Of course if you would mine the dual coin at full potencial, it would affect Ethereum hashrate, that's why we will need to optimize the intensity of the dual coin; lowering it at such degree that it's not affecting Ethereum hashrate.

Dual mining claymore start.bat config:

EthDcrMiner64.exe -epool yourMiningPool -ewal yourEthAddress -epsw x -dpool dualCoinMiningPool -dwal dualCoinWalletAddress -dpsw x -dcri 25

The part before the -dwal is the same as for the solo Ethereum mining described above. The -dwal has the same representation as -ewal, it just is the mining pool of the dual coin. I would recommend to mine ONLY Decred as a dual coin, because it has the highest efficiency of all of them. As described above, Dual coin uses the GPUs core for mining and not all dual coins give the same results. For RX 5xx cards, the best way is to go with Decred. I use the Supernova Decred mining pool. You need to create an account there, and the account name will serve you as a Decred mining pool address. This way, it gives you one more security improvement, because you don't show people your address, instead just your account name. On your account, you will need to create a worker and give it a name for example: worker1, and leave it's password as it is ("password"). Now to connect properly to the Decred mining pool you would need to put

"-dwal supernovaAccountName.supernovaWorkerName"

How to mine Decred?

You can create a Decred wallet at Bittrex. It's a very good trading site featuring a lot of altcoins including Decred. You can cash out your Decred at your account page in supernova, under "My Account" -> "Edit Account" -> "Payment Address" and you need to type your Bittrex address there. And now you just need to set "Automatic Payout Threshold" to your desired value, I use 0.5 as my payout cap. You can convert your mined Decred to Ethereum at Bittrex exchange site, and store the value that way. It's safe if you use a 2FA authenticator.

Important Dual Mining Information:

As you can see in the dual mining configuration, the last part is "-dcri 25". It means that the dual coin is set to mine intensively, and it shows how much GPU core is assigned for that task. Yes, it's needed for solo mining too, and needs to be set to 6. This is a very important part because it's **DEPENDANT ON THE GPU SERIES**. The only noticeable difference between the RX 570 and RX 580 series is their GPU Core. The memory (used for Ethereum mining) is almost the same on those cards, so there is basically no difference in Ethereum hashrate, but the big difference comes in the GPU Core. The RX 580 series can handle around -dcri 25, don't go above that because it can reduce your Ethereum hashrate. For RX 570 series the optimal -dcri is around 19-22. For some cards even lower as 13, this needs to be tested by yourself. The proper way would be to start with -dcri 10. Then using your keyboard press "+" or "-", that way you can increase or decrease -dcri by 1, as you will see on the claymore miner. By going up you will see the dual coin hash rate going up, repeat that until you can start to see the Ethereum hashrate decrease, then, after you find that spot reduce -dcri by 3, so you are not pushing the GPU to the limit. On the RX 570 series, it's possible to get a higher hashrate on Ethereum with dual mining rather than just solo mining. Optimal for RX 570 is around -dcri 19 , optimal for RX 580 series is around -dcri 25. For some cards, it's possible to go even further, but it's not worth it to stress the GPU too much.

Overclock/Undervolt

This is the most important part of this guide, it's very important for you to learn the right way of overclocking and undervolting to optimize the GPU as much as possible.

1. STEP – Delete all the overclock tools that you have installed, especially MSI Afterburner because it can interfere with the proper way of overclocking and undervolting.

2. STEP – Remove any overclock and undervolt from Claymore "start.bat" file if you were using those commands before (-cclock, -mclock, -cvddc, -mvddc) , even -tt and any other option. Claymore overclock and undervolt tools are terrible and have various problems:

- GPUs sometimes randomly reset to a default stock hashrate even if you see the GPU is at its desirable clock rates

- GPUs get random hashrate drops, for example it shows 29.5MH/s, but sometimes it will randomly drop to 20 MH/s for no reason in the logs, and that happens very often. The way Claymore software tries to force your GPUs to stay at those clock rates is terrible, and it will not work properly if a new Driver from AMD comes out.

- You don't have full access to the overclock tools in claymore and it's a mess configuring each GPU individually

- Overclock or undervolt sometimes don't apply properly. Often you will need to start it again 2 times or even restart the PC to have proper Overclock values applied

3. STEP – Reset your GPUs to the default clock rate settings if you have overclocked them before

4. STEP – Download a tool called <u>OverdriveNtool</u>. This is the BEST and most RELIABLE overclock/undervolt tool for AMD graphic cards.

What is OverdriveNtool?

Now after your GPUs are at their default settings, we'll be using OverdriveNtool to handle the overclocking,target the GPUs temperature and its undervolting. There is no other tool where you can have full control of your GPU and the ability to quickly optimize GPUs. You can't be 100% sure the overclock/undervolt settings are working properly. This is a special software that gives you FULL access to your AMD GPUs and it's very easy to use once you know the basics.

How to use OverdriveNtool?

This software may seem confusing or complicated at first, but it's very easy to understand. I will explain it through the following picture:

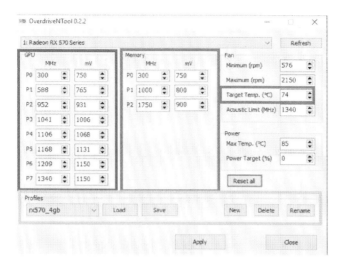

OverdriveNtool Default Settings

Target Temp (small rectangle on the right) – this is the target temperature of your GPU. OverdriveNtool will automatically keep your GPUs at their desired temperature by increasing/decreasing the fan speed, as it's needed to stay at those temperatures. The optimal value would be 60C. You can check this during the mining in Claymore, by seeing how much the current fan speed is in percentage. If the fan's speed goes over 70% increase the target temperature to 65C, but that can only happen if you have a high room temperature, probably because of no cooling or weak air flow.

PROFILES (bottom rectangle) – This serves to save current overclock settings for further use. For example, after you turn on your PC, you can automatically load all the overclock settings to the

desired GPUs. We will **have 1 profile per GPU on your mining rig.**
First make a new .txt file in the folder in which you have
OverdriveNtool.

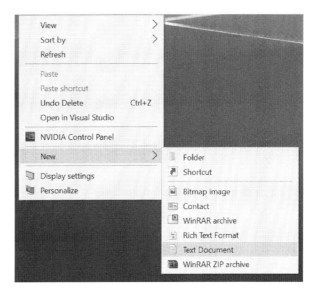

Create .txt File

After that, go to "Save As" and change the "Save as type" to "All
Files" and then name the script "overclock.bat". That way you will
create a Batch file the same type as Claymores "start.bat" and it will
work in very similar way.

Now after that open the overclock.bat file with notepad and write in the following:

OverdriveNTool.exe -r1 -p1"GPU1" -r2 -p2"GPU2" -r3 -p3"GPU3" -r4 -p4"GPU4" -r5 -p5"GPU5" -r6 -p6"GPU6"

As you can see in the following picture:

Overclock Script for OverdriveNtool

This will make a batch script that will run the OverdriveNtool.exe and set each GPU (-p) to a predefined profile ("profileName").

1. -r1 ->resets the GPU 1 to default settings, it's important to always reset GPU before testing new clock settings

2. -p1"GPU1" ->set the GPU 1 values to the profile called "GPU1"

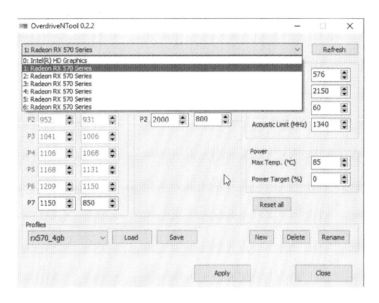

List of AMD GPUs

Be careful, as you can see in the displayed image in my case, there are 7 GPUs enabled on this mining rig. The first one is an integrated GPU and its ID is -p0, all others are mining GPUs (p1,...p6). So if you have your integrated GPU disabled or for some reason you use a motherboard that does not have it, then your mining GPU ID starts from p0. You can see the GPUs order as displayed in the picture below. The GPUs order in OverdriveNtool is the same as in the GPU-Z and Claymore 10.2.

Now make 6 "New" profiles and name them GPU1, GPU2... GPU6 and each profile will represent the GPU it's attached to. For example, we are using "-p1" GPU to the profile "GPU1" and so on.

You need to make so many profiles as you have mining GPUs (all GPUs except the integrated one).

GPU (big rectangle on the left) – this part shows you the real GPU core clock rates and its voltages. In other overclocking tools, you will only see the last one, in this case 1340 MHz. As you noticed, there are 8 of them (P0, P1....P7) and that are the GPUs core states. This means the GPU switches automatically to default between those states, depending on how much you use the GPU. From all those 8 states, we don't want the GPU to switch between them, we want it to run stable at the fixed clock rates we put it on. To do that, we will need to disable all the GPUs states except the last one (P7). You can disable every state from P0 till P7 simply by double clicking on its name. For example, with the mouse go over "P0" and double click on it. You will know if you are successful when that state changes color.

GPU CORE OVERCLOCK/UNDERVOLT – we need to do two things to the GPU core. First, we will need to set P7 clock rate and its voltage. You need to remember that the GPUs core is not used to mine Ethereum a lot, it just helps the memory to do the hashrate. GPU core generates the most heat on the GPU and uses the most power, so our intention is to push the GPU core down as much as possible to save power and lower the temperature on the GPU without losing Ethereum hashrate, or lose some hashrate because we save more on the power cost reduction than the small Ethereum hashrate drop. It is recommended to have Wattmeter to make your own calculations to see what's more worth for you. In general most optimal clock rates for most GPUs is around 1150

MHz. Some RX 570 can even work at around 1100 MHz without losing any, or very low hashrate reduction on Ethereum. That will reduce the power draw drastically. Some RX 580 need 1200MHz to have the optimum hashrate, but most of them work best at 1150MHz. In general, never go above 1200MHz because it will start to use much more power, and you can check that with your Wattmeter. For the Voltage part, it's best to keep them at 850mV. You can try to reduce the voltage to 825mV or 800mV if you are going to keep GPUs at 1100MHz, but it is possible to get a freeze or crash. The best way for you is to test your hashrate with those values described and see what effect it has for your GPU to run it at 1100MHz, 1150MHz, 1200MHz with 850mV voltage in all cases. Then compare the power draw with the hashrate and calculate what's more profitable for you. In most cases 1150MHz/ 850mV is optimal.

MEMORY (middle rectangle) – This works identically as the GPU core, except it's for the memory. This is the Holy Grail, this is the most important part of GPU mining and it's very RANDOM. There is no set values that work 100% on your GPU. There is just one proper way of doing it without risking any problems; we need to disable P0 and P1 by double clicking on them.

How to Set OVERDRIVENTool Properly?

We will need to repeat the process for each GPU individually. It's very important to test it that way, so if you end up getting a crash or reset, you will know exactly at what part that happened so that you can reverse the crashing settings.

First we will need to test the first mining GPU only, not all at once:

1. Set the GPU target temperature to 60C

2. Set the GPU core as explained in the **GPU (big rectangle on the left)** part of the guide (GPU Core) to 1150MHz/850mV, and disable all of the previous states.

3. Set the Memory to 1800MHz/900mV and disable all of the previous states.

4. Apply settings to the GPU.

5. From Profiles find the "GPU1" profile and click Save.

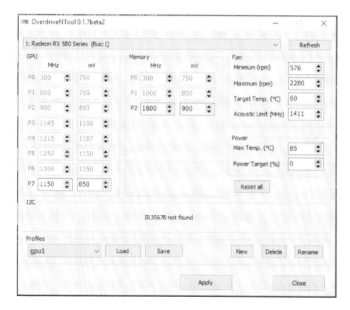

Overclocked and Undervolted Template

As you can see in the picture, you will need to have values set exactly the way it's shown. Apply settings first, then click on the "Save" button near the profile or else the profile settings won't be applied properly. You have your first GPU all set and ready to make the final step.

Overclock Properly

This is the most important question people want to have an answer for and it's the trickiest one. There are no optimal or universal values, because on the identical GPUs, the same Overclock/Undervolt settings don't work the same way. Each GPU is unique and requires individual testing to optimize it properly.

Download a tool called HWinfo64.

Install it and run it in "Sensors only" mode as displayed in this picture:

Sensors only mode for HWinfo64

After that scroll all the way down till you see your GPUs, they are located at the end. Now after you found the GPUs, select all sensors except "Memory Errors" and HIDE them (right click on the sensors and press "hide"). After that, you will have something similar to the image below:

Sensor	Current	Minimum
Read Activity	0.0 %	0.0 %
Write Activity	0.2 %	0.0 %
Total Activity	0.2 %	0.0 %
Read Rate	0.000 MB/s	0.000 MB/s
Write Rate	0.270 MB/s	0.000 MB/s
Read Total	2,085 MB	2,085 MB
Write Total	766 MB	765 MB
GPU [#1]: AMD Radeon RX 580:		
GPU Memory Errors	11	11
GPU [#2]: AMD Radeon RX 580:		
GPU Memory Errors	69	69
GPU [#3]: AMD Radeon RX 580:		
GPU Memory Errors	1	1
GPU [#4]: AMD Radeon RX 580:		
GPU Memory Errors	5	5
GPU [#5]: AMD Radeon RX 580:		
GPU Memory Errors	0	0
GPU [#6]: AMD Radeon RX 580:		
GPU Memory Errors	14	14
Network: RealTek Semiconductor...		
Total DL	0 MB	0 MB
Total UP	0 MB	0 MB
Windows Hardware Errors (WHEA)		
Total Errors	0	0

HWiNFO64 v5.56-3230 Sensor Status [120 values hidden]

0:00:08

Memory Errors in HWinfo64

In my case there are 6 AMD GPUs and I have disabled all other sensors because they are not interesting to me. We only want to have GPU Memory Errors displayed, this will tell you if your GPU is overclocked too much. Now this is the way we will test your GPUs optimal settings.

1. Start mining with the current base settings as we did for first GPU (1150MHz/850mV on GPU core and 1800MHz/900mV) on the Memory, the target temperature is 60C.

2. If you see no memory errors then the current setting is working properly, you will need about 2 minutes to see if they start to show or not.

3. Now change the memory to 1850MHz, then click apply, after that click save on the profile.

4. Now start the overclock.bat script, it will reset the GPU to its default settings, and after that you will need to apply new 1850MHz memory settings.

5. Start mining and see if you get any memory errors after 2-3 minutes.

6. If you don't get any memory errors, that means your GPU is having no problems running at those clock rate. Now we will try to increase memory clock rate by 50MHz increments, and each time you increase memory repeat step 4 and 5. If you see no difference in hashrate after changing memory clock rate, you need to restart your PC. Sometimes if you change overclock/undervolt a lot, it will stop making change. We want to repeat that process till

you start to see memory errors, depending on them do the following:

- If you get few memory errors after some time, go back to 25MHz and test it again.

- If you get millions of errors, that's the HARD CAP of the GPU and you can't push it above that, reduce the clock rate by 50 MHz.

- If you get 1-2 errors each couple of seconds, it is fine and you can keep it like it is.

Now after you have found the optimal value for your GPU, you can do the following:

- Set the GPU core to 1100 MHz and then to 1200 MHz, don't forget to apply, save profile and then run the overclock.bat, to see the difference in hashrate and the power draw from the wall. If you see no difference in hashrate after changing GPU core clock rate, you need to restart your PC; sometimes if you change overclock/undervolt a lot, it will stop making changes.

- Set memory voltage to 850mV and then to 800mV, always look into HWinfo64, because the reduction of voltage on the memory can cause errors. So don't go below 800mV. Reduction in the memory voltage can also cause the GPU to stop working (0 mh/s) or cause the rig to freeze or blue screen. Don't worry about that just change the settings.

After you done all that for the FIRST GPU, you can repeat the process for each other GPUs. Always keep an eye in HWinfo64 for memory errors; that way you will have a stable rig. The rig can work with a bunch of memory errors but that can cause:

- real hashrate to drop

- rejected shares

- seeing random hashrate drops in claymore

- it can cause serious damage to the GPU

If all of your GPUs on the rig are the same, you can try to apply the profile settings that worked for the first GPU to the next GPU and test if it works. After that, try to adjust the small settings to reduce memory errors if you get them. It's possible that the same GPU with the same settings causes the PC to crash or freeze, that's why you test **one GPU at a time.**

Average Optimum GPU Settings

This is a list of different GPUs and their average optimum settings, so that you can have a clear picture of what your GPU should look like:

RX 570 Series:

- **Micron Memory**

  - GPU core: 1150MHz / 850mV

  - memory: 2050MHz / 850mV

- **Hynix Memory**

  - GPU core 1150MHz / 850mV

- memory: 2175MHz / 800mV
- **Samsung Memory**
    - GPU core 1150MHz / 850m
    - memory: 2200MHz / 800mV
- **Elpida Memory**
    - GPU core 1150MHz / 850mV
    - memory 2100MHz / 850mV

RX 580 Series:

- **Micron Memory**
    - GPU core: 1150MHz / 850mV
    - memory: 2050MHz / 850mV
- **Hynix Memory**
    - GPU core 1200MHz / 850m
    - memory: 2250MHz / 850mV
- **Samsung Memory**
    - GPU core 1200MHz / 850mV
    - memory: 2250MHz / 850mV
- **Elpida Memory**
    - GPU core 1150MHz / 850mV
    - memory 2100MHz / 850mV

Starting to mine at Windows startup

After you've managed to setup all of the GPUs profiles and have tested them with no or minimum errors, you want to make sure that your mining rig works automatically.

1. Create a shortcut to Claymores "start.bat"

2. Create a shortcut to OverdrivenTools "overclock.bat"

3. Press Windows Key + S

4. Type "Run" and press Enter

5. Type in: "shell:startup"

6. A folder will pop up

7. Drag OverdriveNtool and Claymore shortcuts to that folder

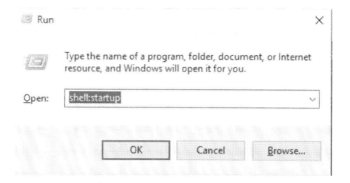

Windows Startup Folder

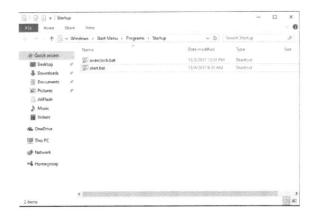

Mining and Overclocking Scripts Starting With Windows

Now try to restart your PC, it should overclock and start mining automatically.

Managing mining rigs

There are many ways of managing your rigs and a lot of software to do that. I like the simplest one, and that works great for me.

1. <u>NANOPOOL</u> – I use this pool to mine Ethereum, and to monitor my rigs. For –ewall, after your Ethereum address, you can put "yourEthWallet.yourRigName/yourReportingEmailAddress". That way, nanopool will send you an email if one of your rigs goes offline. Also at nanopool website you can check your current hashrate of your rigs. If you see one rig reporting lower hashrate than it should you know something is wrong. Nanopool works very well as a monitoring tool.

2. <u>SPLASHTOP</u> – If nanopool finds a problem, the best way to fix the issue is using remote desktop software to access your rigs. Instead of Splashtop you can use Teamviwer they do the same job. And as your Main GPU is integrated one, the streaming software will not have big impact on the hashrate or stability of the rig.

3. <u>SMART PLUG</u> – If the rig is frozen and cannot be accessed with remote desktop, then we need to restart it. I use smart plugs for that. You can monitor your power draw and restart your rigs with it by switching power off and back on.

And that's it, you've optimized your Mining Rig the proper

way.

Hope you've learned a lot and this guide helped you to achieve
a better and efficient mining.

Good luck!

Links to the sites:

.NET Framework 3.5 Offline Installer	https://www.techgainer.com/tools/net-framework-3-5-offline-installer-windows-10-8-x/
Windows 10 Registry tweaks for mining.bat	https://mega.nz/#!8gZzjQaa!Dt4wE0WEo8nZkui_-cAAXL4mb4YlO6CJgFZTXKq9pnQ
DDU	https://www.guru3d.com/files-details/display-driver-uninstaller-download.html
Radeon Software Adrenaline Edition	https://www.amd.com/en/support
Beta Blockchain Driver	https://www.amd.com/en/support/kb/release-notes/rn-rad-win-blockchain-beta
Crimson ReLive Driver	https://www.amd.com/en/support
Pixel Clock Patcher	https://www.monitortests.com/forum/Thread-AMD-ATI-Pixel-Clock-Patcher
GPU-Z	https://www.techpowerup.com/download/techpowerup-gpu-z/
ATIFlash	https://www.techpowerup.com/download/ati-atiflash/
Claymore 10.2 NoFee Version	https://mega.nz/#!h2A0GYJL!OwPnxMRxE-zODBBg9caUFG0Si8Z-AUbUhmoxQhzz990
Bittrex	https://bittrex.com
OverdriveNtool	https://forums.guru3d.com/threads/overdriventool-tool-for-amd-gpus.416116/
HWinfo64	https://www.hwinfo.com/download/

Profitability calculation

If you follow the instruction I'm providing here, it's all very simple; you use your electricity and your computer to generate cash. You may ask where the catch is. Well, it's in a few places; the most obvious is that you have to make sure that the cost you're spending on electricity is less than the value of what you're mining in the currency. Otherwise, you're upside-down and just spending more money on electricity than you're getting back. You can definitely choose to do that if you look at it long term and if you think the value of the coins in future will be worth the cost you're putting into it right now. I strongly believe in the future of cryptos, especially the main ones like Bitcoin or Ethereum. So, even if the prices go down for a while, it doesn't scare me. But, it's your decision if you want to invest in it or not.

There are profitability calculators that determine the profit you can make. This is just going to be a quick and dirty profitability calculation for any of the GPUs in your system.

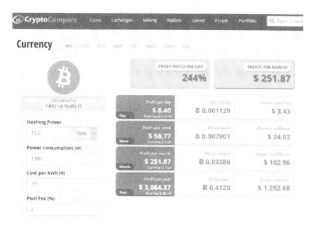

CryptoCompare: a calculator website

You just need to keep in mind, there are two major variables that these calculators typically don't take into account because they're very difficult to predict. One is the future price of the currency and the other is the difficulty increases with mining. Considering these variables, it's actually really tough to predict what the exact time of return on investment will be for your setup.

There's definitely value in getting into cryptocurrency mining right now for many people; however, it could just as quickly hurt you if you invest in it and expect to make a quick return and things go down. My aim isn't to tell you what to do but rather give you information to determine what works best for you.

If you have any questions regarding hardware requirements or how to set up your system, please contact me, and I will do my best to help you. My contact information is at the end of this book.

Conclusion

Cryptocurrency and blockchain technology are here to stay regardless the rise or drop in the price of Bitcoin or other altcoins. This technology has managed to gain great popularity during the span of only a decade. Cryptocurrencies have provided an astonishing opportunity for investors to increase their wealth. So, learning more about this new movement and educating ourselves will help in making better decisions and embracing the opportunities when they come in future. If you don't want to be left out, then make use of the information in this book and start to get involved, either by investing or even mining.

Bitcoin is an innovative technology; there's still more to be improved and changed. Trading and investing in this space can become stressful, and it can be a really lonely thing. My advice to you is to join a Bitcoin or trading community. It is always helpful to be around other like-minded people that can support you. You can also learn from them or teach them the knowledge and experience you have gained. This way you can grow faster and minimize the risks compared to working as an isolated investor.

Last but not least, thank you for reading this book! If you enjoyed it or found it useful, I'd be very grateful if you'd post a short review on the Amazon website. Your support really does make a big difference and I read all the reviews personally, so I can get your feedback and make this book even better. It will also help other people to make an informed decision about my book.

For more information, please check out my website www.futurentech.com.

Other books by me:

The Future and IoT: Building the Internet of Things

Here's the link to the book on Amazon.com

https://www.amazon.com/gp/product/B07587W61D

Thanks again for your support!

Printed in Great Britain
by Amazon